윤한진 한승재 한양규
전보림 이승환
심희준 박수정

Hanjin Yoon Seungjae Han Yangkyu Han
Borim Jun Seunghwan Lee
Heejun Sim Sujeong Park

젊은 건축가
질색, 불만 그리고 일상

윤한진 한승재 한양규
전보림 이승환
심희준 박수정

펴낸날 ┊ 2019년 11월 20일 초판 발행
2020년 3월 25일 2쇄 발행

글쓴이 ┊ 윤한진 한승재 한양규 전보림 이승환 심희준
박수정 정지돈 김재관 조남호
펴낸이 ┊ 김옥철
주간 ┊ 문지숙
기획편집 ┊ 박성진 윤솔희
진행 ┊ 임선희
편집 도움 ┊ 서하나 박지선
번역 ┊ 노성자
영문감수 ┊ 노성화
디자인 ┊ 오혜진
사진 ┊ 김용관 노경 석준기 신경섭 이남선 임준영
전영호 정정호 PLUS 202 Studio
제작 도움 ┊ 박민수
커뮤니케이션 ┊ 이지은 박현수
영업관리 ┊ 한창숙

인쇄 ┊ 스크린그래픽
제책 ┊ SM북

펴낸곳 ┊ (주)안그라픽스
우10881 경기도 파주시 회동길 125-15
전화 031.955.7766(편집) 031.955.7755(고객서비스)
팩스 ┊ 031.955.7744
이메일 ┊ agdesign@ag.co.kr
웹사이트 ┊ www.agbook.co.kr
등록번호 ┊ 제2-236(1975. 7. 7.)

제12회 젊은건축가상 2019
주최 ┊ 문화체육관광부
주관 ┊ 새건축사협의회 한국건축가협회 한국여성건축가협회
후원 ┊ 국민체육진흥공단

이 도서의 국립중앙도서관 출판예정도서목록(CIP)은
서지정보 유통지원시스템 홈페이지(seoji.nl.go.kr)와
국가자료공동목록시스템 (www.nl.go.kr/kolisnet)에서
이용하실 수 있습니다.

CIP제어번호 ┊ CIP2019044224
ISBN 978.89.7059.022.6 (93540)

본 사업은 국민체육진흥기금 후원을 받아
시행하는 사업입니다.

2

Young Architect
Loathing, Dissatisfaction and the Everyday

Hanjin Yoon, Seungjae Han, Yangkyu Han
Borim Jun, Seunghwan Lee
Heejun Sim, Sujeong Park

Publication Date ┊ November 20, 2019
Second Edition ┊ March 25, 2020

Authors ┊ Hanjin Yoon, Seungjae Han,
Yangkyu Han, Borim Jun, Seunghwan Lee,
Heejun Sim, Sujeong Park, Jidon Jung,
Jaekwan Kim, Namho Cho

President ┊ Okchyul Kim
Chief Editor ┊ Jisook Moon
Planning & Editing ┊ Sungjin Park,
Solhee Yoon
Assistant ┊ Sunhee Lim
Proofreading ┊ Hana Sur, Jisun Park
Translation ┊ Seongja Ro
English Proofreading: Seonghwa Ro
Design ┊ Hezin O
Photograph ┊ Yongkwan Kim, Kyung Roh,
Joonki Seok, Kyungsub Shin, Namsun Lee,
Juneyoung Lim, Youngho Chun, Jungho Jung,
PLUS 202 Studio
Production support ┊ Minsu Park
Communication ┊ Jieun Lee, Hyunsue Park
Customer Service ┊ Changsuk Han

Printing ┊ Screen Graphic
Binding ┊ SM Book

Publisher ┊ Ahn Graphics Ltd.
125-15, Hoedong-gil, Paju-si, Gyeonggi-do
10881, Korea
Tel ┊ +82.31.955.7755
Fax ┊ +82.31.955.7744
Email ┊ agdesign@ag.co.kr
Website ┊ www.agbook.co.kr

The 12th Korean Young Architect Award 2019
Presidented by the Ministry of
Culture Sports and Tourism
Organized by Korea Architects Institute,
Korean Institute of Architects,
Korean Institute of Female Architects
Sponsored by Korea Sports Promotion Foundation

A CIP catalogue record for this book is
available from the National Library of Korea,
Seoul, Republic of Korea.

CIP Code ┊ 2019044224
ISBN 978.89.7059.022.6 (93540)

# 젊은 건축가
## 질색, 불만 그리고 일상

# Young Architect
## Loathing, Dissatisfaction and the Everyday

안그라픽스

4

Contents

FHHH friends

프롤로그

지나치게 솔직하고
무섭도록 강직하고
사소하지만 진솔한
: 박성진

Prologue

Excessively Honest
Formidably Strong
Commonplace
but Earnest
: Sungjin Park

박성진
책임에디터, 사이트앤페이지 디렉터

건축가가 쓰는 (혹은 건축가를 소개하는) 책 대다수가 작품 중심의 서사 구조다. 작품을 A, B, C, D 병렬적으로 나열하면서 개념과 상황, 의도 등을 설명하는 게 일반적이다. 똑같은 여건과 배경에서 지어지는 건축은 세상에 없기에 개개의 작품이 놓인 대지 상황과 사회 인문학적 맥락, 또 건축주들의 파란만장한 사연까지 쉽게 글의 소재가 되곤 한다. 이렇게 작품을 설계하면서 체득한 경험적 상황들과 특수해特殊解를 글로 풀어 쓰는 게 비교적 수월하다.

그동안 '젊은건축가상' 단행본도 이 구조에서 크게 벗어나지 않았다. 하지만 2019년은 좀 다르다. 개개의 작품보다 그 작품들을 가로지르는 건축가의 사유와 집착이 무엇인지 그리고 배경에서 작동하는 공통의 문제의식 등이 전면에 부각되었다. 수상자 세 팀은 다섯 개의 주제어를 앞세워 작품을 자연스럽게 해체하고 각자의 사유를 설명하는 재료로

Sungjin Park
Editor, Director of Site&Page

The majority of books written by architects (or about architects) form a narrative around projects. The norm is to explain the concept, situation, and intention by listing the projects as A, B, C and D, line by line. As architecture is never constructed under identical conditions or in identical settings, the situation of the site, the socio-human context that the project is subject to, the wild client anecdotes all easily become fodder for the content of the architectural text. As such, it is a relatively easy task to write up an article detailing the experiential circumstances and specific solutions.

Past Korean Young Architect Award publications have not diverged far from this format. However, this year 2019 is a little different. Rather than focusing on individual projects, the book brings to the fore the architect's thought processes and preoccupations across the spectrum of their work, and taking a step back, the issues that they shared an awareness of. Armed with five keywords, each of the three nominated teams organically dismantle

재구성했다. 이렇게 걸러지고 추려지고
압축된 다섯 개의 주제어는 수상자들이
갖고 있는 습관 혹은 집착이고 아니면
문제의식이거나 주제 의식이며 이도 아니면
늘 잠재적으로 작동하는 배경 논리일 것이다.
올해 수상자들은 자신의 작품을 그럴싸한
비주얼로 포장하려는 욕망과 의도적 거리를
두면서 작품에 대입한 사고들을 추출해 다른
이야기들로 재구성해갔다.

　　이런 구성은 일단 출판 기획자인
나의 제안과 실험이었지만, 동시에 올해
수상자들이 갖는 이 책에 대한 요구와
기대였다. 보통은 건축 작품집의 모습을
기대하며 많은 양의 사진과 상세한 도면으로
작품을 일일이 자세하게 소개하길 원하지만,
올해 그들은 이런 책의 모습을 거부하며
사진보다는 글로, 작품보다는 건축적 사유로
책을 채워가길 원했다. 내 경험상 건축가들이
이렇게 작품화와 이미지 현혹에서
벗어나기란 결코 쉬운 일은 아니다.

이런 상황에서 추출된 다섯 가지 주제어는
그래서 각 건축가가 오랜 건축적 사유와 실천
속에서 충돌하고 깨지고 다듬어지며 무겁게
남겨진 이슈들이다. 이 주제어들은 그들
건축의 출발점이기도 하고, 때론 배경이며
아니면 하나의 결과와 지향점으로 읽혀진다.
하지만 흥미로운 점은 이 세 팀이 전혀 다른
색깔을 갖고 있음에도, 우리 도시를 대하는
관점에서 어느 정도 공통분모가 읽힌다는
것이다. 늘 아웅다웅 싸우듯 즐기듯 논쟁하는
'푸하하하프렌즈'의 관계와 협력이 글에 녹아
있고, 사회를 향해 거침없이 쓴소리를 내뱉던
'아이디알'이 왜 그래야 했는지 알 수 있으며,
건축을 통해 일상의 지평을 넓혀가려는
'건축공방'의 노력을 엿볼 수 있다. 그래서 이
책의 제목 '질색, 불만 그리고 일상'은 하나의
이야기인 듯 세 이야기를 보여준다. 지나치게
솔직하고 무섭도록 강직하고 사소하지만
진솔한 그들의 이야기를 시작한다.

their projects, as a source to
explain their thought processes.
These filtered, whittled down, and
compressed five keywords are the
daily habits or obsessions of the
winners, their problem-solving
and subject awareness, and the
workings of their latent theoretical
understanding. This year's winners
intentionally distanced themselves
from the desire to embellish their
projects with flashy visuals, instead
reconstructing narratives with the
thoughts derived from their work.
　　While I, as an editor, had
first proposed and experimented with
this idea, it was also requested
and hoped for by this year's young
architects. In general, architectural
portfolios are assumed to be a
detailed account of each and every
project, backed up by a myriad of
photos and detailed drawings. Yet
this year, the architects rejected
such a book, instead seeking to fill
its contents with text rather than
image, architectural thoughts rather
than projects. In my experience, it
is no easy task for architects to
detach themselves from the allure of

showcasing their projects and images.
　　The five keywords derived in
this way are what remains from each
architect's struggle and conflict,
weighty issues that have weathered
breaking and reshaping. These
keywords can be read variously as
a point of departure, a setting,
a single accomplishment or the
objective of their architecture.
Notably, despite the contrasting
character of the three teams, the
methods they use to approach the city
do possess commonalities. The ties
and partnership of FHHH friends, as
they continue to quarrel, jostle,
and jest permeates their writing,
we can understand the audacious yet
bitter critique directed to society
by IDR, and appreciate the efforts of
ArchiWorkshop to widen the horizons
of the everyday through architecture.
Hence, the title of this book 'Young
Architect: Loathing, Dissatisfaction,
and the Everyday' is simultaneously
a single story and three separate
stories. Their excessively honest,
formidably strong, commonplace but
earnest stories begin here.

# 푸하하하 프렌즈

# FHHH friends

푸하하하프렌즈는 윤한진, 한승재, 한양규 세 명의 소장과
여섯 명의 동료들로 구성된 건축설계사무소다. 윤한진,
한승재, 한양규 셋은 디자인캠프 문박 디엠피에서 만나
동료로서 인연을 맺었으며 2013년부터 현재에 이르기까지
도시에 대한 폭넓은 이해를 바탕으로 독창적인 작업을
보여주고 있다. '2019년 젊은건축가상'과 '2019 올해의
주목할 만한 건축가'로 선정되었다.

FHHH friends is an architectural design office
consisting of three directors, Hanjin Yoon, Seungjae
Han, Yangkyu Han and six colleagues. The directors
met at design camp moonpark dmp partners and worked
together since 2013 creating an unique project based
on a broad understanding of the city. They are
nominated the 2019 Korean Young Architect Award.

# 어쨌든 프렌즈

# Nevertheless friends

Essay One

#축구 #골키퍼 #공격수 #파트너십

ↄ 어떠한 결의나 굳건한 다짐은
없었다. 다만, 구성원 각자의
삶을 지켜가며 함께 한발을 더
내딛어보는 믿음과 신뢰를 바탕으로
푸하하하프렌즈만의 시간을
쌓아가는 중이다.

푸하하하프렌즈는 함께 일하는 세 친구의 우정과
파트너 관계로 맺어졌지만, 프로젝트를 진행할
때만큼은 각자의 영역이 매우 뚜렷하다. 『젊은 건축가:
질색, 불만 그리고 일상』에 실릴 이 글을 숙제로
받았을 때도 우리의 색깔을 어느 한 명이 도맡아 쓴
글로 규정해도 되는지, 또 그것이 가능하기는 한지를
놓고 고민에 빠졌다. 한 팀으로 같은 방향을 지향하는
것은 분명하니 저마다 각자의 말을 꺼내놓는 것도
적절하지 않았다. 그래서 우리의 뚜렷한 개성을
하나의 글로 엮어줄 수 있는 제3자 김상호에게 도움을
요청했다. 정신이 아득해지는 인터뷰 속에서 건진
이야기들을 통해 되도록 있는 그대로 우리 모습을
전달하고자 했다.

푸하하하프렌즈는 일을 할 때 성취해야 할 목표를 두지
않는다. 거의 놀듯이 일한다. 그렇다고 일을 신나게 하는 건
아니다. 일할 때 드러나는 성향에 대해 잠깐 이야기해보자.
축구 경기로 치면 승재와 한진은 열심히 뛰어다니는

#Soccer #Goalkeeper #Striker #Partnership

We first became acquainted as friends and
partners, but when working we each have
a particular role. When we received this
assignment, our first instinct was to wonder
if one voice could authentically represent
all our voices, or whether that was even
possible. The fact that we are a unified
team and clearly in pursuit of the same
direction made it seem inappropriate for each
of us to write our own separate story. So we
requested the help of an outsider who could
combine our three unique styles into a single
story. We have attempted to communicate our
true selves as much as possible through the
stories salvaged from this daunting interview
process.

We do not really set goals when we are working.
Work usually feels like play. Which doesn't
necessarily mean that the thought of work makes our
heart beat faster with excitement. Let's talk for
a moment about each of our tendencies that come
to the fore when we work. If we were to compare
ourselves to football players, Seungjae and Hanjin
could be seen as players who are busy scouring
the field. Seungjae is a lot like a graceful but
lazy Junghwan Ahn, never leaving the goalposts of
the other team, and not really descending down

↗ They haven't really
resolved anything, nor
are they desperate to say
together. However, based on
faith and trust, they are
building up their time.

플레이어 느낌이다. 승재는 상대 골문 앞에서 벗어나지 않고
후방으로는 잘 내려오지 않는, 우아하지만 게으른 안정환
같다. 한진은 공만 쫓아다니는 것이 영락없이 최용수다. 일단
공을 잡으면 패스할 생각도 없이 마구 내달린다. 반칙도
상관없다. 골만 넣으면 된다. 양규는 골키퍼 김병지 같다.
뒤에서 상대 골을 막으면서 시끄럽게 작전 지시를 하다가
느닷없이 한 번씩 뛰어나가서 슛을 쏘기도 한다. 우리는 지적
능력이나 기술보다는 성격으로 서로를 보완한다. 처음 일을
같이 시작했을 때는 우리가 그럴싸한 일을 할 수 있을 거로
생각하지 못했다. 그냥 같이 놀아보자는 정도였다. 그런데
어느새 한진은 우리 팀이 계속 앞으로 나아가는 원동력이
되었고, 양규는 일을 추진하고 실행하는 데 불가능은
없다는 인식을 심어주었다. 승재는 우리 팀에서 감수성을
담당한다. 우리에게는 한계가 없고, 또 못할 것도 없다. 지금
푸하하하프렌즈에는 그런 시너지와 열정이 있다.
　　살면서 서로 존중하면서도 마음 편한 인간관계를
맺기는 쉽지 않은데, 우리가 그런 사이인 것 같다. 사실 워낙
많이 싸워서 이제 서로에게 필요한 적당한 거리를 알기
때문이기도 하다. 우리는 어떠한 결의도 하지 않고, 서로를
절실하게 원하지도 않는다. 다만 우리 사이에 시간이 쌓인 것

to the rear. Hanjin is unmistakably Yongsu Choi,
always chasing the ball. Once the ball finds him,
he doesn't even consider passing it on, he just
keeps running. It doesn't matter if he cheats, as
long as he scores. Yangkyu is like the goal-keeper,
Byungji Kim. He lies in defense, blocking out
goals, shouting out strategies for the team, then
every now and again he'll suddenly surge forward
and shoot. We complement each other through our
personalities rather than our intellectual capacity
or technique. When we first started working
together we didn't even think we might be able to
make something special. It was more of an attempt
to have some fun. But over time, Hanjin became
the engine of our team, constantly pushing us
further on, while Yanggyu implanted the conviction
that nothing is impossible with the right drive
and execution. Seungjae specializes in being
empathetic. We don't have any limits, or anything
we can't do. For FHHH friends, at this moment in
time, we have both synergy and passion.
　　It's not easy to find people with whom you
feel comfortable, but can also respect, and that's
what it's like for us. To be honest, we've fought
so much that we already know the right amount of
distance to be kept from each other. We haven't
really resolved anything, nor are we desperate
to stay together. It's more about how much time
has passed. Now, it's more about thinking that
it would be nice to keep going together. And the

같다. 이제는 계속 같이 있었으면 좋겠다는 생각만 들 정도다.
그럴수록 서로 조금 더 존중해야 한다. 물론 그런데도 화가
나면 바로 욕이 나오는 어쩔 수 없는 구제 불능들이기도
하다. 아마도 이런 부분이 다른 건축사사무소와의 차이를
만드는 것이 아닐까 싶다. 우리는 끝까지 같이 잘해보자는
사업적 목표도 없고, 친구 사이의 의리 같은 것도 그리
중시하지 않는다. 우리의 모습을 생각했을 때 적당히 각자의
삶을 지켜가다가 끊어지기 직전에 다시 이어붙이는, 자꾸만
그런 모습이 떠오르는 걸 보면 그것이 푸하하하프렌즈의
파트너십인 것 같다.

more we think that, the more we should respect
each other. Of course, there are always those
inevitable moments of rage induced cursing. Perhaps
this is how we're different from other teams. We
have no driving business strategy, nor do we feel
constrained by or obliged to maintain a sense of
loyalty between friends. Each of us more or less
gets on with our own life, and at those moments
when it seems like we might just drift apart we
always end up reconnecting. The fact that it's
those moments that come to mind suggests that
that's the form of our partnership.

스페이스 깨 (2016)
서울시 종로구 채부동에 위치한 ‹스페이스 깨›.
동네에 흔한 빌라 건물을 갤러리로 리노베이션하며
평범한 풍경을 존경하는 자세에 대해 고민한 프로젝트다.

SPACE KKAE (2016)
<Space kkae> located in Chaebu-dong, Jongno-
gu, Seoul. It is a project that considers the
attitude of respecting ordinary scenery while
renovating a typical villa building.

# 질색하고 남은 것

# What remains after loathing

Essay Two

#서울 #아버지 #콜렉티보커피로스터스 #다가구주택
#성수연방 #공장 #리모델링

↘ 서울시 신청사의 측면 모습.
한 나라의 수도에 새로 세워지는
중요한 건물임에도 여러 상황 때문에
건축가의 의도대로 구현되지 못한
모습에 마음이 아팠다.

질색은 애증에서 나온다. 아버지를 떠올릴 때 진저리나게
싫은 구석이 있지만 마냥 미워할 수 없는 존재라고 느끼는
것과 같다. 해외 건축가가 한국에서 중요한 프로젝트를
맡았을 때 건축계에서 불만이 나오는 모습을 종종 본다.
우리는 거기에 별로 개의치 않는다. 그저 잘 지어진 건물에
목마를 뿐이다. 한 나라의 수도에 새로 세워진 시청 건물을
보고 할 말을 잃었더랬다. 외국 건축가든 한국 건축가든 누가
짓든 질 좋고 예쁜 건물이 하나라도 더 지어졌으면 좋겠다.
아버지가 어디서 바가지를 쓰고 비싼 옷을 사 입었더라도
그게 아버지에게 어울리고 멋지면 좋겠다. 서울을 보면 그런
감정이 든다. 태어나서 자랐고 떠나본 적 없는 이 도시를
우리 셋은 매일 같이 돌아다녔다. 어떤 '척하는' 건물들, 겉만
번지르르하게 해놓은 것들이 눈에 띌 때마다 짜증이 났다.
'내 스타일 아니네.' 하고 그냥 넘길 수도 있었지만, 그게
잘되지 않았다. 반드시 짚고 넘어가야 할 문제라고 느꼈다.
그러다 보니 프로젝트에서도 단순히 우리만의 디자인 방식을
추구하는 그 이상의 어떤 책임감이 늘 따라다녔다.

#Seoul #fathers #ColectivoCoffeeRoasters
#MultipleUnitHousingBlocks #Seong-Su-Yeon-Bang
#factory #remodeling

Intense dislike originates from a love and hate
relationship. For examples, your father might
inspire intense feelings of dislike, but that's
not to say that all you feel for him is hate. It's
quite common for the Korean architectural scene to
complain when a foreign architect comes to Korea
and takes on an important project. This doesn't
really bother us. We are just want to see good
architecture. We were once struck dumb by the new
city hall in the capital of Korea. Whether they be
a foreign architect or domestic architect, whoever
it might be, our only wish is that another good
quality, well designed building be constructed.
Whether or not your father has bought an outfit at
an outrageous price, the important thing is whether
it fits him well and makes him look good. Seoul
makes us feel this way. The three of us would daily
wander around this city where we were born, and
grew up, and never left. It was irritating whenever
we saw a building pretending to be "something,"
a building that was all shiny on the outside. It
would have been possible for us to just overlook
it, and feel satisfied with saying "it's not
our style." But we just couldn't. It felt like
something that we had to point out, and sort out
before moving on. That's when we started to feel

↗ A side view of the New
Seoul City Hall. Even
though the building is of
significant value, it has
not been implemented as the
architect intended due to
various circumstances.

가장 질색하는 것 가운데 하나는 간단한 방식으로 다
덮어버리는 것이다. 예를 들면 건물 이음새를 꼼꼼하게
작업하지 않았으면서 완성도를 높이겠다고 몰딩을 붙여 대충
가려버리는 따위다. 노동과 정성이 필요한 작업을 간단히
무마하는 잔머리다. 당연히 수반되어야 할 계획이 무시된
채 최종 이미지만 흉내 내는 것을 매우 싫어한다. 시트지가
싫고, 치장 벽돌 타일이 싫다. '질색'이라는 말은 지금 우리
작업에 필요한 태도를 만들어준 중요한 원동력이다. 다른
건축가처럼 이런 생각도 한다. '이번에는 이런 건축적 실험을
해볼까?' '이런 특이한 형태를 만들어볼까?' '희한한
땅이 있으니 이상한 것을 해볼까?' 하지만 그 이면에는
'내 이번엔 기필코'라는 한국인 특유의 응어리가 있다.
'나는 아버지처럼은 안 살 거야.'라는 생각이 우리를 다른
건축가들과 다르게 만든다.

　　질색의 힘에서 출발한 대표 작업이 서울 용산구
한남동 카페 ‹콜렉티보 커피 로스터스Colectivo Coffee Rosters›다.
기존 건물은 주변에서 흔히 볼 수 있는 다가구 주택, 즉
빌라였다. 한남동은 주거 지역이지만, 빠르게 상업 지역으로
변해가고 있었다. 이 건물이 어떠한 모습으로 변해야 할지
많이 생각했다. 유행하는 식(우린 이런 것이 질색이다)으로

this nagging sense of responsibility in our work,
something beyond just being responsible to promote
a design method unique to ourselves.

    What we dislike most is the use of simple
solutions as a cover up. For example, sometimes
the joints of a building aren't up to standard, so
molding has been used to try and hide away shoddy
work. This is more than a trick, an attempt to
use simple means to cover up what would otherwise
demand a labor of love. We really dislike things
that just pretends to be what it should be in
actuality, while in fact ignoring the necessary,
requisite. We don't like sheet covers or ornamental
brick tiles. "Loathing" is a term which has become
an important motivation for the work we do today.
While we do, like other architects wonder whether
we should "try out this kind of architectural
experiment?" or "try making this kind of peculiar
form?" or "Try doing something bizarre on this
strange site?" on the inside, there is always
something of a native Korean angst which whispers
"whatever it is, we're going to make it happen".
What sets us apart from other architects is
the conviction that "we will not live like our
fathers".

    One of the representative projects powered by
our loathing is the ‹Colectivo Coffee Roasters›.
The existing building was surrounded by generic
multiple unit housing blocks. Hannam-dong is a
residential area that is quickly transforming into

콜렉티보 커피 로스터스 (2017)
골목길에서 본 측면 모습.
기존 창문을 그대로 유지한 채 디자인을 다듬었다.

COLLECTIVO COFFEE ROSTERS (2017)
A side view from an alleyway.
The architects designed the side wall while
keeping the existing windows intact.

하면, 외벽에 까만 페인트를 칠하고 그럴싸한 네온사인을 달고 과장된 싸구려 장식을 걸어둘 수도 있다. '난 원래 빌라였어. 근데 난 그 모습 그대로 커피숍이 되기로 했어. 난 쿨 하니까.' 이런 천박함이 싫었다. 우리는 건물의 원래 모습을 최대한 유지하면서 다음 단계로 넘어갈 수 있는 미래지향적 방법을 고민했다. 그러자 눈에 들어오는 것이 벽돌뿐이었다. 대단히 특별할 필요는 없다고 생각했다. 지금까지는 외부만 벽돌로 되어 있었다면, 이제는 아예 안팎 전체가 벽돌로 된 건물을 상상해봤다.

　요즘 오래된 주택을 리모델링할 때 중간 과정은 다 건너뛰고 마지막에 뭔가 '인더스트리얼 스타일industrial style'처럼 흉내 내거나 빈티지한 느낌을 내기 위해 건물 일부를 일부러 허물고 뜯는 방식이 유행이다. 이는 대부분 공간 설계는 뒷전으로 하고 재료 선택에만 매달리기 때문에 나타나는 현상이다. 여기에는 공간을 어떤 방식으로 마무리할지에 대한 고민과 결정의 과정이 먼저 수반되어야 한다. 그리고 그것이 마지막에 재료 사용 방식으로 연결되어야 한다. 우리는 기존 건물 외벽은 그대로 남겨두고 필요한 부분에만 새 벽돌을 추가로 쌓아 마감했다. 외관 재료를 통해 이 건물이 상업 공간임을 확실히 드러냈다.

a commercial district. We thought a lot about how this building should transform. If we were to follow the trendy route, which we loathe, we would paint the project in black, hang a stylish neon sign and some over the top kitsch decorations. "I was originally a villa. But, I have decided to become a café, in exactly the same shape that I was when I was a house. Because I am cool". We hated this crudeness. We decided to think about a future-centric way in which the building could retain its original shape, while also going to a different level. Ultimately, what came to mind were the bricks. We didn't think uniqueness was a requisite for the project. If in its past life, only the exterior has been covered in brick, we imagined a building in which the entirety of its interior would also be covered in brick.

　There is this current trend for those remodeling old houses, where all the middle processes are skipped out, and they simply demolish a part of the building to imitate a certain "industrial" style of building and give it a "vintage" sensibility. This is a phenomenon common to those who don't prioritize spatial design and obsess over selecting materials. It is important to carefully consider and decide how the space will be completed and through what means. There must be follow through from the selected method and the material that is used at the end. We left the outer wall of the existing building exactly the

↗ The building before the renovation <Colloectivo Coffee Rosters>. The building was one of the generic multiple-unit housing type.

24

콜렉티보 커피 로스터스 (2017)
입구 모습. 주위를 다소 산만하게 만들던 상층부
창문을 벽돌로 말끔히 메꾸고 입구만 열었다.

COLLECTIVO COFFEE ROSTERS (2017)
The building's entrance. The upper window,
was filled with bricks, and only the gate
stayed open.

콜렉티보 커피 로스터스 (2017)
실내 모습. 붉은색 벽돌과 흰색 천장의 조화가 돋보인다.

COLLECTIVO COFFEE ROSTERS (2017)
A view of the interior. The combination of
red bricks and white ceiling stands out.

그리고 원래 있던 돌출창을 살려두어 건물의 옛 모습을
짐작할 수 있는 단서를 두었다. 단열 문제를 해결하는
일이 번거로웠지만, 흔적을 남기는 것이 중요했다.
건물 내벽을 대부분 철거하면서 구조체를 건물 외벽 쪽으로
몰고 외장재인 벽돌을 실내까지 끌어들여 바닥과 벽을 모두
벽돌로 감쌌다. 그 안쪽에 필요한 구조 보강을 하고 단열재를
시공했다. 이중으로 두꺼워진 벽돌벽은 육중한 덩어리 감을
갖게 되었다. 우리는 일관된 재료 사용을 통해 이 건물에
굳건한 정체성을 부여했다.

서울 성동구 성수동의 복합쇼핑시설 〈성수연방〉도
우리의 질색에서 시작한 작업이다. 우리는 성수동의 소문난
곳을 지날 때마다 고개를 저으면서 우리라면 저렇게는 안
할 것이라고 되뇌었다. 그곳을 설계한 사람이나 그런 모습만
즐기는 사람들은 이미 공장을 그저 '빈티지' '인더스트리얼
스타일'로 규정해버린 듯하다. 얼마 전만 해도 성수동에 있던
건물들은 대부분 그대로 철거되었고, 사람들은 그 자리에
뭐가 있었는지 기억조차 하지 못했다. 동네를 지나다니다가
문득 "아, 여기 뭐가 있었는데, 기억나?"라고 물어도 아무도
기억하지 못했다. 그래서 이 성수동 공장 리노베이션 의뢰가
들어왔을 때 잠깐의 고민도 없이 바로 수락했다.

way it was and completed it by adding new bricks
only where it was needed. The exterior building
materials clearly communicated that this building
was a commercial space. We kept the original awning
windows as a hint of what the building had once
looked like. While this was a little inconvenient
for insulation, it was important to leave a trace.
The internal walls were mostly demolished, and
the structure as a whole concentrated around
the building's exterior. The brick, as an outer
material, was drawn into internal spaces, wrapping
the floor and walls with brick. In the meantime,
we worked on the construction necessary for
reinforcing the structure and insulation, The
brick wall, which had doubled in thickness, had
now acquired a solidity and mass. Through the use
of consistent materials, we were able to give the
building its own sturdy identity.

〈Seong-Su-Yeon-Bang〉 is also a project which
was driven by our sense of loathing towards certain
things. Every time we passed by those fashionable
places in Seongsu-dong, we would shake our heads
and remind ourselves that we would never do it
like that if it were up to us. It seems they have
already defined factories as just "vintage" and
"industrial style". It was not so long ago that the
buildings here were simply demolished and people
couldn't even remember what was there before. If
you go past the neighborhood, and ask "hey, what
was there here before? Do you remember?" nobody

성수동의 유행 방식을 따른다면 여기도 답은 뻔했다. 그냥 그대로 두는 것이다. 하지만 그 결정에서 항상 염두에 두어야 하는 것이 실제 운영자의 마음가짐이다. 지금 모습을 정말로 소중히 여겨 그 모습 그대로 가꿔가려는 사람이 아니라면, 건물도 그냥 유행을 좇아 '진정성 있는 척하는' 건물밖에 되지 못한다. 우리는 그 어떤 고민도 하지 않은 채 유행에 동조하기 싫었다.

그리고 또 성수동 공장 지역에서 가장 쉽게 찾아볼 수 있는 리노베이션 방식은 철골 H빔을 이용한 건식 공법이다. 자유로운 철골 구조 안에 원하는 어떤 기능이든 넣을 수 있다. 우리는 건축을 인스턴트처럼 대하는 것이 싫다. 성수동 건축 시장에서는 H빔도 일종의 인스턴트다. 성수동의 모든 공장이 다 H빔으로 증축되고 장식되면 아무 존재감 없는 건물들만 계속 늘어날 뿐이다. 물론 쉽게 떼었다 붙였다 할 수 있는 것은 건식 공법의 장점이다. '유연성'이 좋은 말이기도 하지만, '고민 안 함'의 동의어라고 우리는 생각한다. '내가 결정해줄 수 없으니 당신이 알아서 자유롭게 쓰면 돼.' 하고 책임을 떠넘기는 꼴이다.

상업 공간에는 그곳만의 인상이 필요하다. 그래서 도시와 마주하는 건물의 자세가 중요하다. 우리가 선택한

remembers. That's why when a client inquired if we would do a Seongsu-dong factory renovation we said yes straight away, without a single moment of hesitation.

Again, there was a very clear trend in Seongsu-dong. We could have just let it be. But what must always be considered in this decision is the attitude of the actual owner. If it is not somebody who really wants to preserve the building exactly as it is because it is so precious to them, it would end up becoming nothing more than a building which "pretends" to be authentic in order to follow a trend. We didn't want to become a part of this trend without giving it more thought.

In addition to this, the remodeling method that stands out the most in the commercial spaces of Seongsu-dong are the projects using dry construction methods with steel H beams. It's possible to put whatever functions one wants within the free steel structure. We detested the way that architecture was being treated like an instant object. In the architectural market of Seongsu-dong, the H beam has also become a sort of instant code. If all the factories in Seongsu-dong were to be extended with H beams and decorated accordingly, there would just be more of the same buildings that have no sense of self at all. Of course, an advantage of the dry construction method is that it's possible to easily add things and take them off again. While flexibility is good, we

↗ The building before the renovation <SEONG-SU-YEON-BANG>. The form in which a courtyard sits at the center, with the two buildings on the side facing each other at a little distance, and the sensitivity of the space still remains.

성수연방 (2018)
건물 중정 모습.

SEONG-SU-YEON-BANG (2018)
A view of the courtyard.

성수연방 (2018)
중정에서 바라본 건물의 정면.

SEONG-SU-YEON-BANG (2018)
A facade view from the courtyard.

방법은 건물에서 절대 (웬만해서는) 수정이 불가능한
요소인 기둥을 전면에 내세우고 그 뒤로 사람들이 이용할
만한 '그늘과 길의 공간'을 구획하는 것이었다. 기존의 건물
기둥에 새로 기둥을 덧대었지만 건축적으로 아무 의미가
없다. 일종의 조각품이다. 그 덕분에 리노베이션한
〈성수연방〉을 보면 예전 모습은 떠올릴 수 없다. 그렇다고
옛날 모습이 완전히 사라진 건 아니다. 중정을 사이에 두고
양쪽 건물이 특정한 거리에서 마주 보는 형식과 그 공간의
감수성은 그대로 남아 있다. 우리가 유지하고자 한 것은 바로
그 구도와 공간이 주는 느낌이다. 〈성수연방〉 작업은 상업
공간을 바라보는 일반적인 시선에 대해 우리가 내민 다른
답이다. 많은 사람이 우리가 만든 것을 보고 어딘가 다르다고
생각해 그로써 무작정 유행만 좇아가는 것을 멈췄으면 한다.

also felt that it was being used as a synonym for
"not caring". It felt as if the responsibility was
passed on by saying "I can't decide so you can just
use it however you want to".

   All commercial spaces need to make its own
impression. That's why the frontal gesture of the
building is so important. The method we decided on
was to uncover the columns which were elements of
the building that could almost never be modified in
the building, and behind these create a shady space
for people to use as a path. The new column that
we added to the existing building's columns had
absolutely no function architecturally. It was a
sort of decoration. If you see how ‹Seong-Su-Yeon-
Bang› has been renovated you can't even imagine
what it used to look like. That doesn't mean to say
that its former self has entirely disappeared. The
form in which a courtyard sits at the center, with
the two buildings on the side facing each other
at a little distance, and the sensitivity of the
space still remains. What we wanted to maintain
was the composition, and the feeling of the space.
The ‹Seong-Su-Yeon-Bang› project was our proposed
response to the general way that commercial spaces
are seen. We hope that many people can look at what
we made, think about how it's somehow different,
and stop just blindly following trends.

# 소리 없는 기본

# Silent basics

Essay Three

#주택 #세거리집 #채나눔 #ㅁㅁㄷ작은집 #동화마을주택
#디스이즈네버댓사옥

↘ 한양규 소장의 생가 복원도.
푸하하하프렌즈 한승재 소장의
작품이다.

기본에 관해서는 양규의 유년기 경험이 우리 모두에게
영향을 미쳤을지 모른다. 양규는 어린 시절 동산 아래에
살았다. 대지는 평지였지만, 완전히 절벽 밑에 있는 집이었다.
슬레이트 지붕의 오래된 주택이었는데, 태풍이라도 불면
집에 있는 통장과 귀중품을 싸 들고 근처 마을회관에 가서
자야 했다. 지붕에 작은 돌멩이가 떨어지는 소리만 들려도
불안했고 마당에 쥐덫을 항상 놓아두었으며 방에는 지네도
돌아다녔다. 어느 날 철거 보상을 받아 집을 이사하면서 철거
현장을 우연히 보게 되었다. 합판으로 마감된 천장을 뜯으니
바로 지붕 슬레이트가 나왔다. 세월이 지나 대학에 진학하고
건축을 하면서 알게 되었지만, 천장에는 당연히 단열재가
있어야 했다. 그제야 생활 환경이 왜 그리 열악했는지 이해가
되었다. 여름에는 너무 더웠고 겨울에는 너무 추워 파카를
입고 잤다. 자다가 얼굴을 내밀면 코끝이 너무 시려 잠이
몽땅 달아났다. 지금은 이 모든 게 마치 우스갯소리 같다.
    건축 작업의 기본은 건물이 건물다운 것이다.
집이라면 마땅히 집으로서 갖춰야 할 기본을 갖춰야 한다.

#Housing #JejuSaegeoriHouse #GrownHouse
#DonghwaVillageHouse #ThisisneverthatHeadOffice

To go back to basics, it's possible that Yangkyu's
childhood has influenced all of us. Yangkyu grew
up at the foot of a small hill when he was little.
The site was flat, but completely under a cliff. It
was an old house with a slated roof, and if at any
moment a typhoon came, they had to take their bank
books and prized possessions to sleep at the nearby
village hall. They would feel insecure anytime
they heard the sound of small pebbles on the roof,
and they always had mouse traps in the garden,
and centipedes in the rooms. One day, they were
given compensation to demolish the building and so
they moved, and by chance he saw the house being
demolished. The torn out plywood ceiling revealed
the slates of the roof straightaway. It was only
after university, and a long time had passed and he
had been working as an architect that he realized
that there should have been some sort of insulation
between the two. It was only then that he
understood why his home had felt so uncomfortable.
In the summer it was too hot and in the winter they
slept in their coats. If his face were to peek out
of his coat while sleeping, his nose would get so
cold that it immediately wake him up. Now, all of
this just seems like a funny story.
    The basics of architecture dictate that a
building should be like a building. When it comes

↗ Picture of house of Yanggyu
Han's birth. Drawing by
Seungjae Han, director of
FHHH friends.

십중팔구 양규 덕분에 기본을 충실히 하는 습관이 우리
모두의 몸에 배었다. 건축주 대부분이 오랫동안 번 돈을
모아 집을 짓겠다고 의뢰해오는데, 전문가라는 사람이
허투루 건물을 지어 판다면? 보지 않아도 화가 난다. 그러나
지금도 버젓이 일어나고 있는 일들이다. 망원동에 사는 한
지인은 신축 빌라인 자기 집 욕실에서 목욕하면서 옆집
아이랑 대화도 할 수 있다고 말한다. 충격적이다. 기본이
제대로 갖춰지지 않은 것이다. 그런데 그 집을 가보면 겉은
번지르르하다. 외벽엔 현무암을 붙였고 지붕은 아연판으로
감쌌으며 지하 주차장은 휘황찬란하다. 그런데 내 집에 앉아
옆집과 대화를 하는 촌극이 벌어진다. 이런 일들이 우리가
기본에 대해 계속 생각하도록 만든다. 기본이라는 것은
어쩌면 최소한의 기준이지만 우리는 지키지 않으면 큰일
나는 일이라고 여긴다.

　　　건축의 기본은 당연히 주택이다. 주택 설계는
사람마다 필요가 다 다르기 때문에 늘 기본값을 새로
설정해야 한다. 우리는 저마다 다른 기본값을 찾기 위해
설문지를 만들었다. 건축주에게 재미 삼아 풀도록 한 뒤 그
결과를 보면 "아, 이번 건축주는 이런 사람이구나." 하고 감을
잡을 수 있다. 또 하나의 긍정적인 효과라면 설문을 통해서

to a house, a house should possess all those basic
qualities that a house should possess. Eight out of
ten times, thanks to Yangkyu thinking hard about
the essential needs, the basic standards required
has become second nature to us. Most clients
commission us with money that has taken them a long
time to earn in order to build a house, and what
would happen if so-called experts were to build
and sell a house that has been built carelessly?
It makes us angry just to about it. But these
are events that take place even as we speak. One
person we know who lives in a newly built villa
in Mangwon-dong said that it's possible for him
to chat through the wall to the kid next door
while taking a bath in his own bathroom. This was
shocking. The building fails on a basic level. When
we paid him a visit, the house was so grandiose.
They had applied basalt rock to the façade and the
roof had been wrapped in zinc, and the underground
parking was magnificent. Yet, it is still the site
of a tragic comedy where you can talk through
the walls to your neighbor while sitting in your
own house. These kinds of events which make us
continuously think about the importance of basic
functionality. What we think of as a basic may only
be a minimum that should be expected, but we feel
that it is standard that must be kept.

　　　Of course, the most basic form in
architecture is a house. Residential design
requires a return to the basics each time, because

건축주가 스스로 집을 짓는 과정을 생각하게 된다는 것이다.
'타일은 무엇으로 할까?' 이런 파편적인 생각을 하던 사람이
설문 문답 이후에는 '나에게 방이 얼마나 필요하지?'처럼
공간 구성과 연관된 질문을 생각하기 시작한다.

현재 주거 공간의 기준은 우리가 생각하는 기본보다
너무 아래에 있다. "자장면 배달 왔습니다." 하고 배달원이
현관문을 열면 집안이 훤히 다 보이는 게 말이 되는가?
다들 그렇게 살고 있으니 으레 그러겠거니 하고 넘기는
것뿐이다. 가족이 함께 야외 활동이라도 다녀오면 샤워도
하고 썼던 물건도 씻어야 하는데 화장실이 작으니 한 명밖에
씻지 못한다. 아기나 강아지도 목욕시켜야 하는데 불가능하다.
가족이 사는 집에서 씻을 공간 자체가 부족하다. 집에서
하늘을 올려다볼 수 없는 것도 당연하게 받아들인다.
예전에는 마당 딸린 집에서는 언제나 하늘을 볼 수 있었다.
하지만 지금은 내 하늘이 없는 것에 아무도 의문을
품지 않는다.

우리는 이런 모든 것이 주거 공간에 기본이 결여되어
생기는 문제라고 생각한다. 방음은 기본 중의 기본이다.
건축가라면 배달 음식이 왔을 때도 사생활을 지켜줄 수 있는
설계 정도는 해야 하지 않을까. 기본만 다시 생각해도 좋은

↘ 번듯한 모양새를 하고 있지만 제
기능을 하지 못하는 주택이 많다.
이처럼 겉은 번지르르하지만 기본을
지키지 못하는 건축물을 보면서
푸하하하프렌즈는 다시금 '기본'을
생각한다.

the needs of its residents can wildly differ. We
each made questionnaires to figure our own set of
basic values. We give it to clients as a way for
them to entertain themselves, and the results often
give us a sense that "aha, this client is this type
of person." Another benefit of the questionnaire is
that the client starts to voluntarily think about
the process of building a house. Someone who would
have thought about detached details like "what kind
of tiles?" starts thinking about spatial questions
such as "how many rooms do I need?" after the
questionnaire.

At the moment, the general standard for
residential spaces lies way below what we would
think of as a basic standard. As a quick example,
should the Chinese food delivery guy be able to
see all the way through your house once he opens
the porch and announces that the food is here? We
simply assume it's ok, because everyone seems to
think it's ok. If a family were to return from
outdoor activities, they should be able to shower
and clean whatever tools they used, but a small
bathroom means that only one person can shower
at a time. It's impossible to bathe a baby or a
puppy. There is a general lack of space to wash
in family homes. We also don't question the fact
that it's not possible to look up at the sky from
our homes. Even though in the past, houses with
madang (courtyards) would always have a view of the
sky. Today, nobody questions the fact that they no

↗ FHHH friends think of the
'basic' again when they see
the building that's messy but
doesn't keep the basic.

설계를 할 수 있다. 그 기본이 무엇인가에 대해 생각해보는 게 중요하다. "집은 그냥 따뜻하면 되죠." 이렇게 말하는 사람은 주택이 갖춰야 할 기본 중의 기본만 생각하는 것이다. 우리는 더 많은 기본을 되찾아주는 일을 해야 한다.

제주시 삼양동에 지은 주택 〈세거리집〉은 기본으로만 채운 집이다. 이 집은 건축주 부부, 아이들, 조부모 3대가 함께 산다. 제주도에서는 특이하게 3대가 모여 사는 집을 안거리(안채)와 두 채의 밖거리(바깥채)로 나눠 짓고, 모든 가족 구성원을 평등하게 대하는 관습이 유독 강하다. 우리는 한 동의 건물 안에 세 채의 집을 넣기로 했다. 그래서 거실도 세 개가 있고, 주방도 아이들 집을 제외하고 세대마다 하나씩 있다. 아이들 집에도 큰 거실과 각자의 방, 테라스가 있다. 건축주 부부 집에도 거실이 있고, 거실을 통해 올라갈 수 있는 부부만의 옥상 정원이 있다. 일반적으로 집을 설계할 때면 하나의 공간을 쪼개 부부와 아이들 영역으로 구분하는데, 이 집에서는 아예 채를 나누어 구성했다. 그것이 본래 집의 기본이라고 생각하고 평등하게 사는 집이라는 마을 속에 각자가 지낼 공간을 평등하게 나눴다.

서울 중랑구 면목동에 지은 〈ㅁㅁㄷ 작은집〉은 매우 작은 집이다. 주어진 공간에서 모든 것을 해결해야

↖ 주택에 대한 사람의 기준치가 저마다 다르므로 푸하하하프렌즈는 건축주의 기본 데이터를 찾기 위한 설문지를 만들었다.

### 푸하하하프렌즈 건축학개론

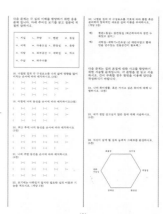

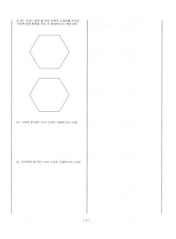

37

longer have their piece of sky.

We believe that all of this is an issue of a lack of standards in residential spaces. Preventing noise pollution should be the most basic of standards. Shouldn't an architect be able, at a bare minimum, to design something which can guard one's privacy against the delivery man? Just rethinking these basic standards can lead to better design. It is important to think about what these basics entail. People who say that a house "just needs to be warm" are only going as far as to think of the very essentials of a house. Our job is to seek out other "essentials" for them.

The ‹Saegeori house› located in Samyang-dong, Jeju is a home filled with such essentials. The house is occupied by three generations. Jeju island has a particularly strong tradition *ahngeori* and *bakgeori*, homes with three generations which divide the home into two annexes, and treats all family members as equals. We decided to place three homes in a single building. Hence, this house has three living rooms, with each unit having their own kitchen except for the children's wing. The children's wing also has a big living room, their own bedrooms and a terrace. The couple's home also has their own living room, and a rooftop garden of their own which they can reach through the living room. Normally, residential design is carried out by splitting up a single space to define the spaces of the couple and the children, but in this house,

↗ Since people's standards for housing differ, FHHH friends have created a questionnaire to find the primary data of the client.

38

세거리집 (2018)
건물 한 동 안에 세 채의 집을 넣었다.

SAEGEORI HOUSE (2018)
The architects decided to place three homes
in a single building.

세거리집 (2018)
마당과 마당 사이에 건축주 부부 내외의 거실이 있다.
아이들은 마음껏 집 안팎을 휘저으며 뛰어다닐 수 있다.

SAEGEORI HOUSE (2018)
Between the yard and the yard, there is a
living room for the client. Their children can
wander freely around indoors and outdoors.

했다. 우리가 가장 신경 쓴 것은 사생활 공간이다. 가족
사이에서도 늘 누군가는 사생활을 침해당하거나 양보하며
지낸다. 사람들은 이것에 익숙해져 있다. 이 문제를 다시
생각해 층마다 마치 다른 세대가 살고 있는 것처럼 설계했다.
가운데 중정은 내부 공간이지만, 햇빛이 내려오는 천창과
노출 콘크리트 계단을 두어 마치 외부 공간처럼 디자인했다.
식당, 거실, 방, 화장실, 안방, 아이 방이 별개의 공간으로
모두 분리되어 있다. 가족이 어디에 있든 간에 각각 분리된
세대처럼 생활할 수 있다.

인천 송월동에 지은 〈동화마을 주택〉은 욕심 때문에
덧붙여진 것들을 덜어내면서 본래 집이 갖춰야 할 기본과
모습을 찾아가는 여정이었다. 마감재를 뜯고 보니 집이
쓰러지기 일보 직전이었다. 1층은 바닥도 없이 보만 덩그러니
드러났다. 그 위에 온수 파이프를 대충 걸치고 시멘트를
얇게 덮고 살았던 것이다. 옛날에는 많은 집이 그랬던 것
같다. 기초도 없었다. 다 뜯었더니 사라진 외부 계단과
아궁이의 흔적도 보였다. 처음에는 보이지 않았던 앞마당도
발굴했다. 이웃 대지를 불법 점유해 증축한 계단도 있었다.
전 주인은 불법 점유로 토지 사용료를 내던 집을 팔 때 대지
면적이 40평이라고 주장했다. 우리는 새 주인에게 집을

the very units were divided in advance. We thought
that this was the basic template of the original
homes. Treating the house like a village, we
divided up the space equitably.

The ‹Grown House› is very small. We needed
to everything to work within the given space. We
were most concerned by the issue of privacy. People
are used to having at least someone within the
family who must give up or yield their privacy.
We rethought this problem. And we designed it so
that it seemed like a different generation lived on
each floor. The central area is an inside space,
but a skylight for sunlight and exposed concrete
stairs give it an outdoor quality. All components,
like the dining room, living room, bedroom,
bathroom, master bedroom and children's rooms are
spatially separate. Wherever the family might be,
it is possible for them to live as if they are in
separate units.

The ‹House at Donghwa Village› was a journey
of erasing what had been added hastily in greed,
and to seek out the basic form of the original
home. When we dismantled the finish, the house was
on the verge of collapsing. The first floor had no
flooring, only beams. They were living in a house
where hot water pipes had been carelessly placed
directly on the beams and covered in a shallow
layer of mortar. In the past, a lot of houses used
to be like this. There were no foundations. Once we
had dismantled everything, we found traces of the

입구 모습. 현관문을 드르륵 하고 열면 숨겨진
넓직한 마당이 나온다.

GROWN HOUSE (2016)
A view of entrance. When you open the front
door, you will see a large yard hidden.

ㅁㅁㄷ 작은집 (2016)
실내 모습. 햇빛이 내려오는 천창과 노출 콘크리트 계단을
두어 마치 외부 공간처럼 느껴지도록 디자인했다.

GROWN HOUSE (2016)
A view of the interior. The skylight and
exposed concrete stairs give us an outdoor
quality.

정상적으로 되돌려놓아야 나중에 다시 매매도 가능하고
집의 가치도 더 좋아진다고 설득했다. 우리는 계단을
과감하게 잘라내며 불법 점유해 증축한 부분들을 하나하나
걷어냈다. 아주 기초적인 구조 설계도 했다. 각 파이프로
만든 전단벽으로 지진에 견딜 수 있는 구조체를 만들고,
그 위에 석고 보드를 붙여 최대한 넓은 공간을 확보했다.
시공사는 진작에 도망갔다. 하지만 우리는 '이 집을 쓸 수
있게 만들어줘야겠다.'는 이상한 사명감에 직접 공사를
마무리했다. 덕분에 집은 단출해졌지만 새 주인은 지금까지
잘 살고 있다.

　　우리는 기본을 건너뛴 채 욕심을 부리지 않는다.
건축을 제대로 배운 사람이라면 모두 마찬가지일 것이다.
최근 우리를 소개한 기사 제목이 '짓기만 하면 핫플…기묘한
건물로 주목받는 건축가들'이었다. 우리 기사는 이런 제목을
달고 나오는 경우가 정말 많다. '실험적인 건축가들', 그런데
우리는 한 번도 실험이라고 생각하고 작업한 적이 없다. '핫플
제조기' '통통 튀는' '종잡을 수 없는' 같은 제목 구절만 보면
마치 우리가 정신 나간 요상한 디자인을 할 것 같은 느낌을
준다. 하지만 이 말들은 우리가 가장 피하고 싶은 말들이다.
아마 우리가 작업을 설명할 때 각자 캐릭터에 빗대어

↗ ‹동화마을 주택› 리노베이션 전 모습.
1968년에 지어진 집으로 불법 증축된
부분이 많았다. 집의 본래 형태로
돌아가는 일이 가장 중요했다.

43

outdoor stairs and the *agungi* (kitchen hearth). We
also excavated the front yard which, at first, was
hidden from sight. There were also stairs which had
been extended by illegally occupying the neighbor's
land. The former owner had asserted that the land
was 132m², including the land he was illegally
occupying. We persuaded the new owners that the
house should be returned to its former state, not
only so that it could be sold in the future, but
also to raise the value of the house. Starting with
the illegal stairs, we started to eliminate the
parts which had been illegally expanded, one by
one. We also carried out a very basic structural
design. We created an earthquake proof structure
with shear walls made of pipes, and directly
attached limescale boards to secure a large as
possible space. The contractors had already run
away. We concluded the construction ourselves, led
by a strange sense that we had a calling to "turn
the house into something useful". As a result,
while the house became a bit more modest, the new
owner still lives there to this day.
　　We don't let greed get in the way of taking
care of the basics. This probably applies to anyone
who has received proper architectural training. A
recent article about us was titled "whatever they
build becomes a hot place…… architects noticed
for their extraordinary buildings". A lot of
articles about us end up with this sort of title,
"experimental architects." Yet we have never

↗ The building before the
renovation of <House at
Donghwa Village>. It was
built in 1968, and there were
many illegal additions to the
building. The architects'
mission was to return to the
original form of the house.

동화마을 주택 (2016)
집의 원래 모습을 되찾고 쓸모 있는 공간을
만들고자 노력했다.

HOUSE AT DONGWHA VILLAGE (2016)
The architects tried to regain the original
appearance of the house and make it a
useful space.

이야기를 하다 보니 그런 현상이 생기는 것 같다. 그것이
푸하하하프렌즈의 딜레마다.

　　도면은 우리 태도를 단적으로 보여준다. 우리 생각을
실제로 구현하겠다는 의지를 보여주는 것이다. 그래서
도면을 그릴 때는 사소한 디테일 하나도 빠트리지 않고
철저하게 그린다. '어떻게든 되겠지.' '나중에 현장 돌아가는
상황을 보고 조치하지 뭐.' 이런 생각은 아예 하지 않는다.
우리가 떠올린 모든 건축적 아이디어는 도면에 반영해야
한다. 도면을 완성했을 때 비로소 우리 스스로 건축주,
시공사, 선후배에게 당당할 수 있다. 그것이 우리가 생각하는
기본이다. 프로젝트 전체 일정에서 가장 집중할 때도 도면
그리는 기간이다. 직원들뿐 아니라 세 소장도 직접 도면을
그린다. 스케치만 해서 넘기지 않는다. 장인이 물건을 만들
때 라이선스만 주지 않는 것처럼 직접 내 손길을 거쳐야만
한다는 생각 때문에 그렇게 한다.

　　실측도 매우 철저하게 한다. 다른 일이 돌아가지
않을 정도로 집착한 적도 있다. 현장을 수십 번 방문하는
건 물론이다. 이제는 '실측 간다'는 이야기만 나와도
직원들이 슬슬 도망 다니는 분위기다. 실측 도면에서는
보통 시작점과 끝점의 실측선이 잘 만나지 않는다. 실제

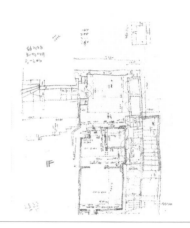

↘ 한양규는 현장 실측을 할 때면
각도를 재고 소수점 셋째 자리까지
확인하며 집요하게 기록한다.

thought of a project as an experiment. All those
adjectives such as "hot place maker" "exuberant"
"unpredictable" are all titles which give you the
impression that we do strange otherworldly design.
The articles all have headings strewn with the
adjectives we would like to avoid the most. This is
probably because we allude to our work through the
characters that get surfaced in our conversations.
For us, this is a dilemma.
　　Our site plans clearly reflect our approach.
They reflect our determination to realize our
thoughts in the real world. So when we are drawing
out the plans, we strict about making sure that not
even the tiniest of details gets left out. We don't
let ourselves think "its going to be ok" "we will
take care of this once we see how things are going
on site". Anything and everything that we think
of architecturally speaking must be reflected in
those site plans. It is only when the site plans
are complete that we can finally feel confident in
front of each other, the client, the contractors
and our seniors and juniors. This is what we
believe to be a basic standard. The most intensive
period of concentration in the overall project
schedule is also designated to the time spent
drawing the plans. The principals and staff alike,
everyone draws the plans. Nobody just hands over a
sketch. This stems from the thought that no artisan
would just hand over his/her license when creating
an artefact, and so our materials should also have

↗ Yanggyu Han measures angles
and checks up to the third
decimal point when doing a
site survey.

건물 벽이 삐뚤삐뚤하니 만날 수가 없다. 그래서 나중에
캐드로 옮기면서 오차를 바로 잡는데 양규가 실측하면 딱
맞아떨어진다. 양규는 각도를 재고, 소수점 셋째 자리까지
확인한다. 실측 도면을 보면 비뚤어지고 튀어나온 선이
있는데 그건 모두 양규가 실측한 각도 때문이다.

　　의류업체 매장이자 사무실인 ‹디스이즈네버댓
Thisisneverthat 사옥›은 기존 주택을 상업 공간과 업무 공간을
겸하는 건물로 변경하는 작업이었다. 기존 주택은 거의
새집이나 다름없었고 나름대로 신경 써서 지은 건물이었다.
'이 멀쩡한 건물을 어떻게 하지?' 이런 생각이 먼저 들었다.
그런데 어딘가 어색해보였다. 한참을 살펴보니 외장재를
사용한 방식이 눈에 걸렸다. 외장재로 콘크리트를 압출해
만든 베이스 패널을 사용했는데 건식 재료를 마치 습식
재료처럼 취급해 설계하고 마무리해놓은 것이 무척 이상했다.
이 건물이 그 자체로 노출 콘크리트 덩어리였다면 이상하지
않았을 것이다. 모서리의 몰딩도 건식 패널을 어색하게
만드는 요소였다. 검은색 석재 타일로 감싸 덧댄 측면도
외장재를 뽐내기 위한 용도였다. 그럴싸하고 쓸모는 있는데
건물과 재료의 어휘가 없었다. 어떻게 하면 이 건물의 어설픈
인상을 바로잡을 수 있을까?

↗ ‹디스이즈네버댓 사옥›의 리노베이션
전 건물 모습. 푸하하하프렌즈는
건물을 처음 보고 어설픈 인상이
느껴졌다고 한다.

gone through our own hands.
　　We are also very strict with our site
surveys. We even had times where we were so
obsessed with it that nothing else was happening.
Of course, this entails visiting the site tens of
times. Now, our staff quietly try to evade us if
we even so much as mention that we will be "going
on a site survey". Normally in site survey plans,
the starting line and the finish line rarely meet
as survey lines. These faults are modified once we
move to the CAD files. There is no way they can
meet since the actual walls of the building are
not straight at all. However, anything we survey,
but in particular by Yangkyu, will be entirely
accurate. Yangkyu measures angles up to the third
decimal point. Sometimes there are wonky lines in
the survey floor plans and these come from the
angles that have been measured on site.
　　For the ‹Thisisneverthat Office Building›
we needed to modify an existing house into a
building with both commercial and office spaces.
The existing house was a new one, and it had been
a house which had been built with quite a lot of
care. Our first thought was "This building seems
decent! What should we do with it?" There was
something that was a little awkward about it, and
after much observation, the method for applying the
finish caught our eye. The façade was made of base
panels produced from compressed concrete, and it
was very strange how the finish gave the impression

↗ The building before
the renovation of
‹Thisisneverthat Office
Building›. The architect felt
the flimsy impression when
they first saw the building.

디스이즈네버댓 사옥 (2019)
세로 방향 파이프를 없애는 방식으로 유리 커튼 월을
설계해 시공했다.

THISISNEVERTHAT OFFICE BUILDING (2019)
In the design, the architects eliminated the
vertical pipes in the curtainwall.

48

디스이즈네버댓 사옥 (2019)
1층 입구 모습. 콘크리트 구조물을 유리
커튼 월이 감싸고 있는 모양새다.

THISISNEVERTHAT OFFICE BUILDING (2019)
A view of the entrance. It seems like
a glass curtain walls are surrounding a
concrete structure.

외부에 건식으로 걸리는 재료는 건물의 껍질이다. 얇지만
단단하고 상징적 힘을 가질 때 비로소 껍질은 파사드가 된다.
유리 커튼 월은 베이스 패널과 대조되는 물성을 갖는다.
단단하고 투명하므로 새로운 공간을 형성하기에 적절하며
베이스 패널의 존재감도 살릴 수 있다. 하지만 일반적인
유리 커튼 월은 기대만큼 개방적이지 않다. 유리를 지지하기
위해 60밀리미터 정도 두께의 파이프로 만든 격자 구조물이
필요하기 때문이다. 안쪽에서는 새장처럼 느껴지기도 한다.
우리는 협력 업체의 도움을 받아 세로 방향 파이프를 없애는
방식으로 유리 커튼 월을 만들기로 했다. 가로 부재로 하중을
받을 수 있도록 했고, 계단 바깥 콘크리트 수벽樹壁과 로드
바Rod-bar로 가로 부재를 지지하도록 했다.

　　파사드가 껍질이라면 계단과 현관 안쪽은 건물의
알맹이다. 알맹이의 콘크리트 피복 두께를 늘려 바깥면에
치핑chipping 기법을 사용해 약간 거칠게 면을 깨고 모서리의
예리한 선은 남겨 전반적으로 무겁고 단단한 물성을 살렸다.
새로운 파사드에는 장식이나 개념적인 이야기 이상의 기능이
있어야 한다. 파사드와 함께 만들어진 측면 계단과 안쪽 벽은
업무 공간의 동선으로 사용된다. 기존 바닥을 철거해 만든
정면 현관은 상업 공간을 위한 출입구가 된다.

that it had been designed with wet materials when
it had actually been designed with dry materials.
It would not have been so strange if the building
had been a mass of exposed concrete from the onset.
The molding finish at the corners was also an
element that made the dry panels look out of place.
The layer of black stone tiles was also intended to
highlight the external materials. It was stylish
and useful, but had no sense of the building or its
materials. We thought hard about how we might set
right these awkward aspects.

　　The dry materials which were hung up were like
a skin. When these skins are thin, yet sturdy, and
have a symbolic significance, they become a façade.
The glass curtainwall has a materiality which
contrasts with the base panel. It is appropriate
for establishing a new space because it is sturdy
and transparent, while simultaneously highlighting
the existence of the base panel. Curtainwalls in
general are not as open as one hopes they will be.
This is because they require a gridded structure to
support the glass using pipes of around 60mm thick.
From the inside, it can feel like a birdcage. In
our design, we eliminated the vertical pipes in
the curtainwall through collaboration with another
company. It was designed to sustain the load with
the transverse member, and so that the concrete
vertical curtain on the outside of the stairs and
the rod bars could sustain the transversal members.

　　If the façade is a skin, the stairs and

the interior of the foyer are the core heart of
the building. We expanded the concrete sheath of
the core, making chipping the concrete to treat
the surface while keeping the corners intact
to preserve its sense The new façade requires
a function which goes beyond ornamentation or
conceptual thought. The new façade has a function
beyond decoration or conceptualizing a story. The
side stairwells and the internal wall which were
made in tandem with the façade is used for the flow
of circulation for the business space. The foyer at
the forefront which was created by demolishing the
existing slabs becomes an exit for the commercial
space.

# 집요함만 남는다

---

# At the end of the day, it's about being stubborn

Essay Four

#어라운드사옥 #삼각형 #성수연방 #흙담 #장인정신

우리는 우리 작업에서 집요함이 보여야 한다는 강박감이
있다. 그것을 이뤄내지 못하면 시간 낭비를 한 기분이 든다.
인생에서 그 시간은 지워버려도 될 것 같다. 집요함은
결국 디테일로 이어진다.

가장 대표 사례가 서울 마포구 연남동에 지은
잡지사 건물 ‹어라운드AROUND 사옥›이다. 세 개의 삼각형
덩어리를 쌓아놓은 형태인데, 삼각형이 하나씩 올라갈
때마다 앞쪽으로 1.2미터씩 튀어나온다. 건물 하부는 햇볕과
비를 피할 수 있는 캐노피가 되고, 건물 후면은 남향 빛을
충분히 받는 테라스가 된다. 이 형태를 구현할 때 삼각형 세
개를 어떻게 쌓아야 우리가 원하는 건물의 인상을 정확하게
전달할 수 있을지 고민했다. 전체 형태를 인지할 수 있도록
건물이 지면에 닿는 부분을 최소화하고 땅에서 띄워 형태에
오롯이 집중할 수 있게 했다. 그리고 1대 1대 0.8의 어긋난
비례로 긴장감을 더했다. 결과적으로 어떻게 보이는가를
강조하고자 했다. 건축 비평가 이종건은 이 작업을 평하면서
"모더니즘은 어떻게 보이는가에 있다."고 했는데, 실제로
우리가 그 부분에 집중한 작업이었다.

#AroundHeadOffice #Triangular #Seong-Su-Yeon-Bang
#GroundWall #ArtisanalSpirit

We are slightly obsessed with whether our projects
reflect a sense of tenacity. If not, we feel like
we have wasted our time. It almost feels like we
could just erase that period of time from our
lives. The issue of tenacity is an issue of detail.
The most representative case is the ‹ Around
Magazine Office Building› at Yeonnam-dong, Mapo-gu,
Seoul. This project is in the form of three stacked
triangular masses, and each stacked triangle
propels the whole project forward by around 1.2m.
The base of the building becomes a canopy providing
shade from the sun and the rain, while the rear
becomes a terrace flooded with sunshine from the
south. We contended a lot with the issue of how
to stack the triangles to clearly communicate the
desired impression. We minimized the area that
actually touched the ground to make the overall
form more discernible, as if it were floating on
the ground, so that the whole form of the project
could fully come into focus. And we added a sense
of tension through the distorted ratios of 1:1:0.8.
We wanted to ultimately bring intentionality to how
it is seen. In reviewing about this project, the
architecture critic Jongkeun Lee said "Modernism is
defined by how it is seen."
Raising the irregularly stacked three
volumes from the ground level, pushes the center

비뚤게 쌓은 세 개의 덩어리를 바닥에서 띄우면 무게 중심이
한쪽으로 치우친다. 구조적으로 매우 까다로운 형태다.
게다가 가장 위에 있는 구조물이 3.6미터 튀어나와 있으면
건물 하중의 절반 이상이 한쪽으로 쏠리게 된다. 벽이나
기둥을 더 늘려 캔틸레버Cantilever 길이를 줄이는 쉬운
방법이 있었지만, 전체 3.6미터의 캔틸레버를 고집했다.
기둥을 세우는 것은 용납할 수 없었다. 형태의 긴장감이
사라지기 때문이다. 결론적으로 가장 아래 삼각형에서
구조의 핵심이 되는 코어를 지하 깊숙이 박아넣고, 2층 바닥
두께를 800밀리미터로 만들었다. 비슷한 규모에서 볼 수
있는 일반적인 바닥 두께의 네 배에 달하는 치수다. 2층
바닥 자체가 거대한 구조체가 되어 전체 건물을 지탱한다.
그 두께 덕분에 엘리베이터도 없는 건물에 계단이 훨씬 더
많아졌지만, 그런데도 건물 사용에 문제가 없도록 하면서
우리 생각을 끝까지 지켜냈다.

　　뾰족한 삼각형 건물을 모형이나 도면으로 그리기는
쉽다. 하지만 실제 땅에 쓸모 있는 3차원 공간이 되도록
만드는 일은 쉽지 않다. 작은 가구 하나도 맞춤 제작해야
한다. 기성품은 뭘 넣어도 모서리에 삼각형 공간이 생긴다.
그런 문제들을 해소하는 것이 이 프로젝트의 또 다른

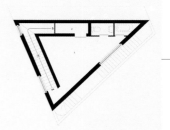

of the project to one side. This is an extremely
particular form in terms of its structure.
Propelling the project forward by 3.6m with this
type of volume leads to more than half of the
project's weight being concentrated on one side.
The easy way out would be to increase the number
of walls or columns to reduce the length of the
cantilever, but we were determined to have a 3.6m
cantilever. We couldn't accept the use of columns.
This would have destroyed any sense of tension
in the form. As a result, we constructed a deep,
underground core inside the structure, and created
800mm thick slabs for the second floor. This is
four times the thickness of a slab generally found
in projects of a similar scale. The second floor
slab itself becomes a huge structure to support
the entire building. To overcome this thickness
we ended up needing more stairs in a building
without an elevator, but nevertheless, we succeeded
in realizing our vision while making sure that
no problems would arise in terms of using the
building.
　　It's quite easy to emulate a pointed
triangular building with models or plans. It's not
so easy to make a useful three dimensional space
in reality. Even the smallest pieces of furniture,
needs to be tailormade. Whatever type of ready-
made product we would put it there, the corners
would create triangular gaps. One of the fiercer/
more intense moments of this project was having to

↗ The shape of the <Around
Magazine Office Building>
looks like a stack of three
triangles.

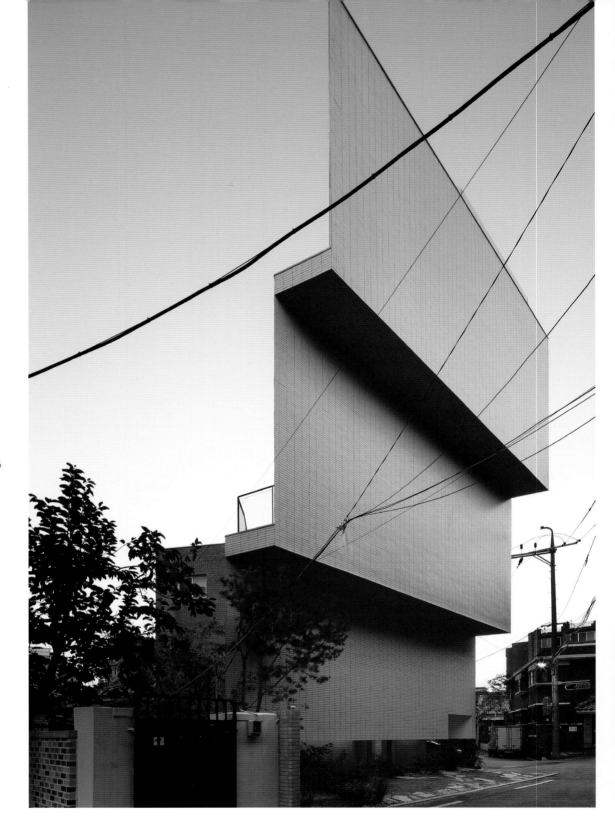

어라운드 사옥 (2017)
세 개의 삼각형 덩어리를 쌓아놓은 형태. 삼각형이 하나씩
올라갈 때마다 앞쪽으로 1,200밀리미터씩 튀어나온다.

AROUND MAGAZINE OFFICE BUILDING (2017)
The form of three stacked triangular masses.
Each stacked triangle propels the whole
project forward by around 1.200mm.

치열함이었다. 실제 사용할 계단 폭을 900밀리미터로
맞추는 것도 보통 일이 아니었다. 도면상 벽체를 포함해
1,200밀리미터가 가용할 수 있는 수치였다. 계단 폭
900밀리미터를 빼고 남는 벽체 300밀리미터에 구조체,
단열재, 타일 마감 두께까지 계산해 넣어야 했다. 특히
계단은 시공 오차가 생기기 쉬우므로 작은 허술함이
쌓이면 전체를 망치는 주범이 된다. 결국 우리는 그 선을
다 지켜냈다. 계단이 건물을 타고 도는 곳곳에서 마치
맞춤옷을 지어 입은 듯한 설계와 시공의 노력이 보일
것이다. 더욱이 우리가 원하는 모양을 만들기 위해 계단의
기본 기능을 희생하지도 않았다. '이것을 위해서 다른
걸 포기했다.'가 아니라 '이것을 위해서 다른 것도 다
충족했다.'가 우리의 치열함이다.

삼각형 세 면이 예각으로 만나는 부분의 타일도 매우
신경 써서 시공했다. 타일로 마감한 입면이 모눈종이 같은
여백으로 보였으면 했기 때문이다. 그래서 벽돌쌓기식이
아니라 정확한 격자형으로 타일을 시공했다. 그리고 각
면의 끝 부분이 예리하게 만나도록 타일 모서리를 30도씩
모따기해 타일과 타일이 빈틈없이 시공되도록 작업했다.
공장에서 가공할 수 있는 한계를 넘어섰기 때문에 타일 깎는

resolve these problems. It was also a tough job to
fit the width of the stairs for everyday use to
900mm. On the drawings, the numbers showed that a
total of 1,200mm was available including the walls.
We had to deduct 900mm for the width of the stairs,
and then calculate everything from the structure,
insulation, and even the thickness of the finishing
tiles to fit into the remaining 300mm walls.
Stairs in particular can easily be miscalculated,
and become a decisive flaw in a project. At the
end of the day, we worked within all of these
rules. One will be able to see that the design
and construction reflects an effort comparable to
bespoke tailoring, particularly in places where
the staircase wraps around the building. Moreover,
we didn't have to sacrifice the original function
of the stairs to realize the form we envisioned.
Our intensity comes from the fact that we "ensured
everything else worked to realize this one element"
rather than the fact that we "sacrificed everything
for this one element".
    We also spent a lot of time considering the
tiling for where the three surfaces of the triangle
meet at an acute angle. We wanted the tile finished
façade to look like clean graph paper. Hence, we
constructed the tiles with an accurate grid form
rather than stacked bricks. We also corner cut
30 degrees each from the edges of each tile to
make the edges of each surface meet sharply with
the next, to work so that there would be no gaps

어라운드 사옥 (2017)
계단이 건물을 타고 도는 모습이다.

AROUND MAGAZINE OFFICE BUILDING (2017)
Notably it looks like the staircase wraps
around the building's shape.

어라운드 사옥 (2017)
각 면의 끝부분이 예리하게 만나도록 타일 모서리를
30도씩 모따기했다.

AROUND MAGAZINE OFFICE BUILDING (2017)
The architects did corner cut 30 degrees each
from the edges of each tile to make the edges
of each surface meet sharply with the next.

어라운드 사옥 (2017)
마치 모눈종이처럼 보이도록 정확한 격자형으로
타일을 시공했다.

AROUND MAGAZINE OFFICE BUILDING (2017)
The architects wanted the tile finished
façade to look like clean graph paper.

칼을 별도로 제작해 맞춤 조정해야 했다.

〈성수연방〉에서 보인 집요함은 앞서 언급한 바깥
기둥을 조각품처럼 만드는 일이었다. 고전적 의미의 조각은
멈춘 시간이라고 생각한다. 조각 작품은 어느 한순간을
공간에 박제하는 것이다. 우리도 이 건물의 파사드가
조각처럼 보였으면 했다. 성수동 건물들에서 이 시대를
박제한 느낌을 받았기 때문이다.

건물 기둥을 조각품으로 만들기 위해서는 배수관
등의 설비를 숨겨야 하는 것은 물론이고, 기둥이 조각품처럼
돋보일 수 있도록 바닥을 얇게 유지해야 한다. 그래서 기존
바닥 두께인 220밀리미터보다 두꺼워지지 않도록 디테일에
집요하게 매달렸다. 오래되어 조금씩 처진 건물 바닥 경사도
일부러 평평하게 교정하지 않고 그대로 따랐다. 두께가
200밀리미터인 기존의 처진 바닥을 반듯하게 덮으려면
새 바닥 두께가 300-400밀리미터가 되어야 하는데, 이를
수용할 수 없었기 때문이다. 이런 고집을 부리느라 시공사와
다투고 험한 소리를 주고받으면서 설계에서 손을 떼려고까지
했다. 우리가 작업하면서 스스로 이런 집요함이 보이지
않으면 또 한 번의 시간을 버리는 느낌이 든다.

김해 진영읍 봉하마을 근처에 지은 복합건물

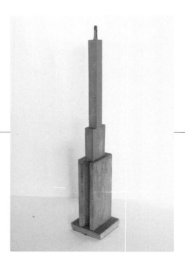

↖ 〈성수연방〉 기둥을 만들기 위한
목업 과정.

between one tile to the next. This surpassed what
could be manufactured in a factory, requiring us to
tailor-make a separate carving knife.

The attention to detail in ‹Seong-Su-Yeon-
Bang› was related to making the outside column like
a work of art. We believe that sculpture in the
classical sense belongs to a frozen moment of time.
A sculpture is an act of preserving a a certain
moment of a space. We also hoped that the façade
of this building could look like a sculpture. This
is because the buildings in Seongsu-dong gave us
the feeling of preserving this period and this
generation.

To make the project's columns sculpture-
like, we not only had to hide utilities like
pipes, we also had to keep the flooring at a
delicate thickness. Therefore, the challenge
was to tenaciously figure out the detail while
ensuring that it would not become thicker than the
220mm found on existing slabs. The gradient of
the project's floor, which was starting to sink
due to age, was also kept the way it was without
deliberately altering it to make it level. A new
floor thickness of 300 to 400 mm was required to
make the existing 200mm slabs level, and it was
impossible to accept this. We even thought of
giving up on the project as we ended up in conflict
with the contractors due to our tenaciousness over
this detail. However, we would also feel like we
put time to waste without such self-enforced high

↗ Mock-up details for the
column of <Seong-Su-Yeon-
Bang>.

성수연방 (2018)
파사드 기둥에 달린 조명 모습.

SEONG-SU-YEON-BANG (2018)
Lighting details on the façade columns.

〈흙담〉은 첫 신축 작업으로, 어쩌면 우리의 집요한 노동의 시작점이었던 것 같다. 언젠가 이 건물이 다 허물어지고 한 가지만 남는다면 우리가 직접 만들어 쌓아 올린 거친 벽돌벽이었으면 좋겠다. 벽돌 디자인의 호불호는 별 의미가 없었다. 지금에 와서 전체 건물을 다시 살펴보면 설계도 잘 못 한 것 같다. 치기 어린 욕심과 욕망 덩어리가 건물 곳곳에 보여서 낯 뜨겁다. 하지만 그곳의 벽돌 한 장 한 장은 지금까지도 정말 특별한 의미로 우리에게 남아 있다. 벽돌을 일일이 손으로 만들어 시공했는데, 작업하는 동안 여러 차례 시행착오를 겪었다. 한꺼번에 여러 장을 만들려고 덤비는 등 바보 같은 짓을 반복하다가 결국 한 장씩 찍어냈다. 건물이 다 사라지더라도 이 장면 하나만은 영원히 기록에 남았으면 좋겠다.

　　푸하하하프렌즈에게는 분명 장인정신이 있다. 끝까지 손으로 만들어내야 하고, 사람 손이 닿는 모든 영역은 우리가 직접 다뤄야 직성이 풀린다. 그래야만 비로소 작업 전체가 잘 끝났다는 생각이 든다. 우리 머릿속 건축가의 모습은 깔끔한 정장에 멋있는 스카프, 동그란 안경을 쓰고 긴 책상에 앉아 스케치하는 사람이 아니다. 현장에서 기고만장하게 서서 지시만 내리는 사람도 아니다. 자기 손으로 직접 뭔가를

〈흙담〉에 쓰인 벽돌을 일일이 손으로 만들어 시공했다.

standards of determination.
　　The first project we designed from scratch, the ‹Ground Wall› might be seen as the starting point for our tenacious labor. If one day, this building has worn down, and there is only one thing left, we hope it will be the rough brick wall which we stacked together. The brick design itself is not very meaningful. Looking back on the entire building today, there are places where we feel the design could have been better. We feel ashamed to see, here and there on the project, a naive greed and ambition. However, each and every brick still remains with us with so much meaning. These were handmade bricks, which went through a tireless process of trial and error during the project. We ended up going round in circles like idiots, trying to make a ton of them at once, and we remember ultimately having to mold them one by one. Even if the entire building were to disappear, we hope that this scene could remain recorded in some way.
　　We definitely have an artisanal spirit. We need to see things through by making them with our hands, and we are only satisfied when everywhere that can be touched by hand has been dealt personally by us. It is only by doing this that we can think that the entire project came to a good end. The image of an architect in our minds has nothing to do with a man wearing a nice suit with a nice scarf and round glasses, sketching at a desk. Nor is it a person found on site yelling

The bricks in the walls of <Ground Wall> were made by the architects.

흙담 (2014)
훗날 이 건물에서 한 가지만 남는다면 직접 만들어올린
이 거친 벽돌벽이길 바랐다.

GROUND WALL (2014)
Later, if there is only one thing left in
the building, the architects hoped it would
be this rough brick wall they built.

만들어나갈 수 있는 사람이다. 그런 면이 집요함으로 나오는
것 같다. 요즘은 점점 남의 손을 빌리는 일이 많아지고
있지만, 여전히 모든 작업에 우리의 집요함의 흔적을
남기려고 노력한다. 〈흙담〉의 직접 만든 벽돌은 우리가
보여준 건축의 장인정신이다.

공식적으로 한 번도 공개하지 않은 푸하하하프렌즈
최초의 작업이 있다. 〈흙담〉과 비슷한 시기에 진행한 집수리
작업인데, 우리가 직접 공사했다. 한 팀은 김해 〈흙담〉 현장에,
한 팀은 서울에서 집수리를 맡았다. 작은 공사라 마땅한
시공 업체도 없었고, 공사비도 얼마 되지 않았다. 〈흙담〉에서
만들고 남은 벽돌을 서울 현장의 담장에 쓰기로 하고 공사를
진행했다.

며칠이 지나 모든 직원이 다 같이 서울 현장의 담장을
보러 갔는데, 입이 떨어지지 않았다. 담장은 너무 엉성했고
전벽돌을 깐 바닥은 높이가 맞지 않았다. 다들 독립한 뒤
경험이 많지 않은 시절이었으므로 누굴 탓할 수도 없었다.
그때나 지금이나 우리 사이에는 '절대 신뢰의 법칙'이라는 게
있는데, 죽이 되든 밥이 되든 '하고 싶으면 하라'는 것이다.
그런데 정작 결과가 그리 처참하게 나오니 아무 말도 할
수 없었다. '그래, 이런 일도 겪어봐야지.' 이렇게 생각하며

at everyone and shouting out orders. It is more
likely someone who can make something with their
own hands. Perhaps it's this dimension that can
be associated with tenacity. Recently, we've seen
a rise in the number of projects we delegate to
others, yet, we still try to leave traces of our
tenacity in all of our projects. We feel that
the bricks in the ‹Ground Wall› reflected this
artisanal spirit of architecture.

Our very first project as FHHH friends has
never officially been disclosed. It was a home
repair project which took place at a similar time
as the ‹Ground Wall›, and we did the construction
ourselves. One team would be onsite at the Kimhae
‹Ground Wall› and the other time managed the
home repairs in Seoul. The small scale of the
construction meant that it was hard to find an
appropriate contractor, and the construction budget
wasn't very high either. We decided to use the
bricks which were made for the ‹Ground Wall› in the
wall, and to go ahead with the construction.

A couple of days later, we went to see
the wall, and it was so terrible that we were
speechless. The wall looked so out of place, and
the floor where the tiled brick had been laid out
did not even meet the level of the project. We
couldn't even blame anyone as it was back in the
day when we were all quite inexperienced, and not
long after becoming independent. Then, as now, we
have always operated on the "principle of absolute

속으로 삭였던 것 같다. 건축주가 하소연을 해왔다. 결국
우리가 받기로 한 비용을 몽땅 들여서 다시 공사했다. 건축주
입장에서는 당장 우리가 아니면 망친 공사를 수습해줄
사람도 없었으므로 울며 겨자 먹기로 마무리를 맡겼다.
재시공을 위한 철거는 간단했다. 이단 옆차기로 툭 찼더니
담장 전체가 와르르 무너졌다. 다행히 두 번째 공사는
잘 마무리되었다. 건축주와도 이후 몇 년 동안 명절마다 안부
인사를 주고받으며 지냈다. 천만다행이었다.

　　　이 작업은 우리에게 무엇보다 좋은 약이 되었다.
실력을 떠나서 현장을 잘 돌봐야 한다는 사실을 뼈저리게
느꼈고, 손해를 보더라도 우리가 만족할 수 있는 것을
만들어야 한다는 개념이 머릿속에 자리 잡았다. 그 뒤로
시공사가 못하겠다고 하면 그냥 우리가 해버리는 경우도
심심찮게 생겼다. 공사하다가 시공사가 도망간 적도 두 번
있었는데, 둘 다 우리가 직접 공사를 마무리했다. 그 이후로
전체 현장 공사가 끝나는 마지막 날이면 우리가 직접 공사를
했든 안 했든 다 같이 가서 우당탕탕 끝내는 전통이 생겼다.
잔손 보기도 다 같이 한다. 모든 현장을 다 그렇게 하진
못하지만 지금도 언제든 함께할 준비가 되어 있다.
일단 시작했으면 끝을 봐야 한다.

trust", which basically means "do what you want
to do" whether it becomes a shambles or not.
Nevertheless, we really didn't know what to say
faced with such disheartening results. We think we
all kept it bottled up inside, thinking "well, this
was always going to be a necessary experience". Of
course, the client complained. In the end, we put
everything we were supposed to earn into re-doing
the construction. The client had no choice but to
entrust us with its completion as there wasn't even
anyone else who was willing to straighten out what
had gone wrong other than us. Demolishing the wall
before reconstruction was easy. One karate kick and
the whole wall came down. Fortunately, the second
attempt went well. We even stayed in contact with
the client for several years after, during the
holidays. It was a big relief.
　　　This served as a good lesson for us. We
were forcibly reminded that looking after the
site was as important as design competence, and
we learnt the lesson that it was better to make
what would satisfy us even if it meant we made a
loss financially. Since then, we've developed a
tradition of all going on site to chime in during
construction, whether we have been involved in it
or not, on the final day. We do everything, even
all the smaller chores. While we can't do this for
all of our projects, we are always prepared to join
in. Once we commit, we see it through.

↗ The first work that
has never been officially
released. As the architects
did the home repairs project,
they felt the importance of
the site.

# 진지함에 대한 알레르기?

# Allergic to seriousness?

#본질 #옹느세자매 #빈브라더스 #에이랜드
#건축평단 #젊은건축가상

지금은 푸하하하프렌즈도 어쩔 수 없이 진지한 말이
필요한 자리에 갈 때가 있는데 우리 캐릭터 때문에 태도가
이중적으로 보인다는 것을 깨달았다. 진지한 '척'에
알레르기 반응이 있었던 것도 사실이고, '우리는 이렇게
재미있어요'라는 콘셉트가 우리에게 있는 것도 사실이다.
설계는 심각하게 하면서 겉으로는 웃긴 척하는 것이 얄밉게
보이기도 할 것 같다. 하지만 다른 사람들이 우리를 어떻게
보는지 별로 중요하지 않다. 그래서 잡지 인터뷰도 늘 편하게
한다. 잡지에서 좋은 의도로 우리를 소개한다고 하니 편하게
이야기하는 것이다. 일부러 건축에 대한 올곧은 태도를
내보이려고 하지도 않는다. 우리 셋은 태생적으로 성격이
밝고, 아직은 에너지도 넘치기 때문에 사람들에게 산만하게
떠드는 모습이 더 각인되는 것 같다. 하지만 그건 우리가
건물을 대하는 태도와는 별개의 문제다.

　　우리는 우리가 한 만큼에 관해 이야기하고 평가받는
게 중요하다고 생각한다. 그래서 결과물에 비해 과도한
미사여구를 붙이는 걸 몹시 싫어한다. 예를 들면 "부수고

↳ 왼쪽부터 차례대로
푸하하하프렌즈의 윤한진, 한양규,
한승재 소장.

#Essence #OnNeSaitJamais #BeansBrothers #ALAND
#OpenArchitecture #KoreanYoungArchitectAward

Now, we have inevitably started to go to places
where we have to talk about serious things, but we
realized that our characters often mean that our
attitude can look quite two faced. It's true that
we felt allergic towards anything that attempted to
"be serious" and it's also true that we had a kind
of concept to show people that "we are so fun".
It's doesn't feel like a stretch to think that
they must hate us for always pretending to be funny
while taking design so seriously on the inside. But
it doesn't really matter to us what we look like
to others. That's why we have always felt quite
relaxed about interviews with magazines. We just
felt comfortable talking because the magazine said
they would present us with good-will. We didn't try
to present some sort of ideological righteousness
about architecture. The three of us are just
naturally quite lighthearted, and we still have
more energy than we can handle, and I think that
has kind of imprinted itself in people's minds,
the sight of us chatting away noisily. But that is
a separate issue from the attitude we have when it
comes to buildings.

　　We believe that it is important to talk about
and be evaluated for all the work we have done.
Hence, we really dislike people who frequently
oversell their work with over the top adjectives.

↗ From the left, Hanjin Yoon,
Yanggyu Han, and Seungjae
Han, principals of FHHH
friends.

원래대로 했어요."라고 설명하지 "본질에 대한 탐구가 이번 프로젝트의 시작이었습니다."라고 하지 않는다. 소탈하거나 웃기려고 하는 것이 아니라 그냥 딱 우리가 한 만큼만 말하려고 하기 때문이다. 일하면서 우리끼리 평가를 주고받다 보니 생긴 태도 같기도 하다. 누군가 자기 생각을 포장하려고 들면 곧장 그러지 말라고 지적한다. 말보다는 결과로 보여주어야 하고 우리는 우리가 한 것만 이야기한다.

　'어렵게 하지 말자.'라는 생각도 있다. 설계 자체를 배배 꼬아서 하지 말자는 의미다. 특히 상업 건축에서 건축가의 의미 부여가 과해지면 거기에서 오는 이질감이 더 도드라진다. "두 재료가 만나는 부분을 이런 몰딩으로 처리한 이유는 두 세계의 만남을 뜻하는 것으로……." 작업을 마땅히 설명할 말이 없을 때 의미를 더하려다 보니 이런 실수가 생기는 것이다. 돈 벌려고 한 작업에 그런 의미까지 넣으려니 너무 속물 같다. 그렇다고 "그냥 했어요."라고 성의 없이 말하지는 않는다. 피부에 와닿을 수 있는 건축을 하자는 생각을 늘 한다. '와닿는다'는 의미는 '본질을 꿰뚫는다'는 말과 통한다. 어떤 프로젝트를 할 때 한 사람이 그려온 그림을 보고 와닿지 않으면 그때부터 다른 둘이 그 프로젝트가 끝날 때까지 놀려대기 일쑤다.

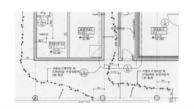

↘ 아직 굳지 않은 시멘트 바닥을 유유히 밟고 사라진 고양이의 흔적을 도면으로 기록했다.

For example, we prefer to say, "After getting rid of A, we worked on B to return it to its former state." rather than saying "the departure of this project stemmed from an investigation into authenticity". It's not that we are humble or that we want to make people laugh, it's just an approach where we prefer to only explain as far as what we did when describing a project. I think it might also be an attitude which came out of working together and criticizing each other's work. Anytime someone tried to package what they were thinking, we would call them on it. Things should be seen in the results rather than in words, and we only want to discuss what we have done.

　We've also thought that we shouldn't make things too difficult to understand. This basically means that the design itself shouldn't be too hard to understand. Especially in commercial architecture, you can feel a sense of distance when the architect attaches too much meaning to his work. Mistakes happen when you try to add meaning to a project that doesn't necessarily have much substance. Basically, explanations like "we finished such and such molding over the part where two different materials meet, as a means to signify the meeting of two worlds." It seems a little superficial to try to impart meaning like that to a project which was clearly done to make money. That's not to say that we should respond frivolously and carelessly and just say "We just

↗ A drawing shows the traces of a cat who disappeared gently on the cement floor before hardening.

빈브라더스 (2016)
바리스타와 손님이 바에 나란히 있는 모습을
연상해 설계했다.

BEAN BROTHERS (2016)
The architects went a step further to
visualize a scene in which the barista and
the visitor are sitting together by the bar.

서울 용산구 한남동 카페 ‹옹느세자매 On ne sait jamais›를 예로
들어보자. 건축주에게 내가 학교 운동장 스탠드에 앉았을
때 경험을 이야기했다. 계단식 스탠드에 앉아 있으면 친구가
뒤에서 말도 걸고, 마주 걸어오는 아이들에게 인사도 하면서
인간관계가 이루어지는 모습이 재미있다고 생각했기
때문이다. 그게 설계의 시작이자 끝이었고, ‹옹느세자매›를
그런 공간으로 만들겠다는 목표로 디테일을 다듬어나갔다.
카페의 전형적인 모습이 있다면 우리는 거기에서 한 번 더
고민해본다. “이렇게 편하게 앉아 있어도 되지 않을까?”
이런 질문은 ‘카페가 이런 모습일 수도 있지 않겠는가?’ 하고
화두를 던지는 셈이다. 그다음부터 우리 이야기를 조금씩
덧붙여나간다. 그러면 설계하기도 설명하기도 훨씬 쉬워진다.

　　경기도 하남의 복합쇼핑센터 스타필드에 있는 카페
‹빈브라더스 Bean Brothers› 작업에서 삼각형 배치를 생각하게
된 계기도 커피숍의 전형적 배치에서 든 의문 때문이었다.
커피숍에 가면 바 뒤에 바리스타가 있고 맞은편에 손님
자리가 있다. 카페는 커피를 만드는 사람과 마시는 사람
사이에 즐거운 소통이 이뤄지는 게 그 공간의 즐거움이라고
생각했다. 그래서 한 걸음 더 나아가 기존의 바리스타-바-
손님의 구조가 아니라 바리스타와 손님이 바에 나란히 있는

ↆ 카페 ‹빈브라더스›에서는 커피를
만드는 사람과 마시는 사람 사이에
즐거운 소통이 일어나길 바랐다.

73

did.” We have always thought about architecture
that you can feel through the skin. Feeling through
the skin is somehow the same thing as saying “being
transparent in its origins”. If one of us were to
draw something out and nobody felt anything about
it, then the other two will certainly make fun of
that person until the project comes to an end.
　　For example, with the ‹On ne sait jamais›
coffee shop located in Hannam-dong, Seoul, I told
the client about my experience sitting on the
school steps. If you sit on the school steps at
school, then friends will come up to you from
behind and start talking to you, and the children
who are walking towards you will say hi, and we
thought it was fun to see how that's how new
friendships are formed. That was the beginning and
the end of the design, and we started to work on
the details with the objective to create a space
like that. We thought about it again when thinking
about the café as an archetype. Questions like
“is it ok to be sitting so comfortably?” felt as
if they were posing the question, “cafes could
look like this too”. After that, we just needed
to develop our thoughts a little more. Then, it
becomes a lot easier to design and to explain.
　　The reason we came up with a triangular
arrangement for ‹Beans Brothers› located in
Starfield Hanam also stemmed from our question
about the typical arrangement found in coffee
shops. If you go into a coffee shop, the barista is

↗ The architects wanted to
create a space where occurs
natural communication between
the barista and customers.

에이랜드 (2016)
매장 가구 디테일.

ALAND (2016)
The furniture detail.

모습을 연상해 설계했다. 같은 쇼핑센터에 있는 의류매장 〈에이랜드ALAND〉에서도 쇼핑이라는 행위를 한 번 더 생각해 그냥 쓱 보고 나가는 게 아니라 오랫동안 그 공간을 산책하듯이 돌아다니면 좋겠다고 생각했다. 그래서 네 개의 공간으로 골목길을 만들어서 사람들이 그냥 통과하지 않도록 설계했다.

진지한 건축에 대한 알레르기는 없다. 다만 기존 어휘에 대한 알레르기는 있다. 우리가 지금 쓰는 말, '커피란 무엇일까?' '어떻게 앉아야 할까?'는 정확히 알맹이를 건드린다. 그런데 건축에는 건축을 표현하는 알 수 없는 많은 어휘가 존재한다. 우리는 프레젠테이션용으로 습관처럼 쓰는 그런 단어들을 거부한다. 우리가 스스로 이해하지 못 하는 말을 누가 알아들을 수 있을까? 건축하는 사람이라면 당연히 설계에 많은 생각을 함축한다. 우리는 그런 생각들을 그냥 쉽게 말하는 편이다. 나름대로 진지하게 이야기하지만 거기에 미사여구가 없으니 무심하게 들리는 게 아닐까.

우리의 동기는 오직 건물이다. 도면도 아니고 모형도 아니고 스토리도 아니고, 오로지 건물로 말한다. 이 건물을 잘 짓고 싶다는 생각만큼은 확실하므로 이걸 누가 망치려 하면 싸우고 건축법규를 샅샅이 검토하며 공무원을

behind the bar, and opposite you have the places for the visitors to sit. We thought that the pleasure derived from this space was created by the entertaining communication that takes place between the person who makes the coffee and the person who drinks the coffee. Hence, we went a step further to visualize a scene in which the barista and the visitor are sitting together by the bar, rather than the composition of barista, bar and visitor. In 〈ALAND〉 we also though that it would be nice if people were to stroll around the space for a long time, rather than just quickly glancing at things and leaving. So, we designed it so that the four volumes would act like alleyways so that people wouldn't just pass through.

We are not really allergic towards serious architecture. However, we are allergic to the current language. The language we use today "What is coffee" "how should we sit" hits the core. Yet, we reject the countless concepts that we don't understand that are used to express architecture, or the terms that are habitually used for presentations. Who would be able to understand anything that even we ourselves don't understand? Any architect is bound to evoke many different thoughts into a single design. We prefer to just talk about it as it is. We sometimes talk about it seriously among ourselves, but it might be the lack of adjectives that makes it sounds as if we don't care.

에이랜드 (2016)
네 개의 공간으로 골목길을 만들어서 사람들이
그냥 통과하지 않도록 설계했다.

ALAND (2016)
The architects designed it so that the
four volumes would act like alleyways so
that people wouldn't just pass through.

찾아다닌다. 이런 것들에 신경을 곤두세우고 지어지는 모든
것에 민감하게 반응한다. 그런데 그 외의 것에 대해서는
성취동기가 거의 없다. 건물을 짓는 일 외에 다른 분야의
일에서 유익을 찾을 때 우리 세 사람이 그걸 어리바리하게
준비하는 모습을 보고 직원들이 웃는 게 좋다. 우리는 늘
직원을 웃기고 싶다. 어쩌다가 잘 되어서 무슨 상금이라도
조금 떨어지면 같이 맛있는 거 먹는 정도로 생각한다.

　　'2019 젊은건축가상 수상'이라는 타이틀은 탐낼
만하다. 작년에 쓴 잔을 한 번 마시고 나서야 사람들이 이
상을 가볍게 생각하지 않는다는 것을 알았다. 상의 의미를
깨닫고 나니 곱씹을수록 탐이 났다. 효도도 할 수 있고
누구에게든 '건축가' '아키텍트Architect'라고 당당히 말할 수
있다. 그동안 공항 입국신고서의 직업란에 승재는 '사업가',
한진은 '자영업자'라고 썼다. 이제 '2019 젊은건축가상'이라는
이름에 '건축가'라고 적혀 있으니 앞으로는 '건축가'라고 써도
된다. 그런 목마름이 우리에게 있었다는 것을 뒤늦게 알았다.

　　예전에 계간 잡지 《건축평단》에서 섭외 연락을 받은
적이 있다. 제대로 된 건축계의 평가를 받아본다는 사실에
너무 설렜다. 그동안 우리에게 관심을 보였던 곳은 대부분
인테리어 잡지였는데, 《건축평단》 같은 데서 대놓고

Our only motivation is the project. It's not the
plan or the model or the story, we speak solely
with the building. What is clear to us, is that
we would like to build a project well, and hence
if anyone tries to destroy it then we fight, we
look up all the laws and visit the officials. It's
this sort of work that really grabs our attention.
We respond very sensitively to everything that is
built. But we are not really driven by any sense of
motivation apart from this. Other than designing
buildings, what we enjoy most is making our staff
laugh at how clumsy and awkward the three of us
are. We always want to make our staff laugh. If we
are lucky enough to be awarded some prize money, we
take it as a good reason to go out have a good meal
together.
　　The title of the Korea Young Architect Award
is something to covet. It was only after we were
sorely disappointed last year that we understood
that people do not take this award lightly.
Perhaps it made us want it even more, the more we
reconsidered after realizing the significance of
the award. It was a chance to make our parents
proud, and to be able to say confidently to anyone
out there that we are architects. Until now, on the
arrivals declaration form on the flight back home,
Sungjae would put "businessman" and Hanjin would
put "self-entrepreneur". Since it says "architect"
in the title Young Architect's Award, we can just
say that we are architects from now on. It's only

'너희 작품 다뤄 주마'라고 하니 말이다. 우리를 진지하게
바라보는 사람들이 생겼구나 싶었다. 같은 맥락에서 '2019
젊은건축가상'도 우리에게 의미가 크다. 우리가 그 세계에
관심 없는 척했던 것은 '누가 우리를 알아주겠어.' 하는
마음이 있었기 때문일 것이다. 이제 돌아보니 제도권에서
우리를 가만 놔두지 않았다는 느낌도 든다. 우리가 가야 할
방향이 제도권을 피하기에는 접점이 점점 많아지고 있다.

*    이 글은 『젊은 건축가: 질색, 불만 그리고 일상』을
     위해 푸하하하프렌즈와 2019년 7월 30일에 인터뷰를
     진행하고 재구성해 인터뷰이의 에세이 형식으로
     편집한 내용이다. (인터뷰·편집 김상호, 녹취록 정리
     심미선)

↘ 건축을 향한 푸하하하프렌즈의
동기는 오직 '건물'을 향하고 있다.

after that we belatedly realized that we had a
thirst for that kind of thing.
    When we were first invited a forum by
«Open Architecture», we were really excited that
we would be receiving proper feedback from the
architectural field. Until now, the majority of
outlets interested in us were mostly light interior
design magazines, and now a place like «Open
Architecture» was willingly coming forth to sat,
"we will willingly feature your work". We felt that
now there were people who would take us seriously.
In the same context, the Korean Young Architect
Award is highly meaningful to us. Perhaps we were
pretending nonchalance, was because we felt that
there wouldn't be anyone who would think anything
of us. Now that we look back, it kind of feels like
the field never really left us alone. It is also
true that the direction we are taking increasingly
has too many points of contact to successfully
avoid the field.

*    This text is an edited version of an
     interview carried out with FHHH friends on
     the 30th of July 2019 into the form of an
     essay of the interviewees for the purpose of
     this book (interviewer·editing article Sangho
     Kim, editing transcript Jane Misun Shim)

FHHH friends.
Their only motivation is
designing a building.

설계 개요

동화마을 주택(리노베이션)
· 설계 ┆ 푸하하하프렌즈(윤한진, 한승재, 한양규)
· 위치 ┆ 인천시 중구 송월동3가
· 용도 ┆ 단독주택
· 대지면적 ┆ 77m²
· 건축면적 ┆ 40.69m²
· 연면적 ┆ 69.19m²
· 규모 ┆ 지상 2층
· 높이 ┆ 7.7m
· 주차 ┆ 0대
· 건폐율 ┆ 52.84%
· 용적률 ┆ 89.86%
· 구조 ┆ 연와조
· 외부마감 ┆ 외단열 시스템
· 내부마감 ┆ 석고보드 위 수성페인트
· 구조설계 ┆ 터구조
· 시공 ┆ 건축주 직영
· 설계기간 ┆ 2015. 3. - 11.
· 시공기간 ┆ 2015. 12. - 2016. 7.
· 사진 ┆ PLUS 202 Studio
· 건축주 ┆ 황성욱

Design overview

HOUSE AT DONGHWA VILLAGE (RENOVATION)
· Architect ┆ FHHH friends (Hanjin Yoon, Seungjae Han, Yangkyu Han)
· Location ┆ Songwol-dong 3ga, Jung-gu, Incheon, Korea
· Programme ┆ housing
· Site area ┆ 77m²
· Building area ┆ 40.69m²
· Gross floor area ┆ 69.19m²
· Building scope ┆ 2F
· Height ┆ 7.7m
· Parking capacity ┆ 0
· Building coverage ┆ 52.84%
· Floor area ratio ┆ 89.86%
· Structure ┆ brick structure
· Exterior finishing ┆ Outer insulation
· Interior finishing ┆ painting on gypsum board
· Structural engineer ┆ Teo Structure
· Construction ┆ managing client
· Design period ┆ Mar. - Nov. 2015
· Construction period ┆ Dec. 2015 - July 2016
· Photograph ┆ PLUS 202 Studio
· Client ┆ Sungwook Hwang

ㅁㅁㄷ 작은집
- 설계 ┆ 푸하하하프렌즈(윤한진, 한승재, 한양규)
- 위치 ┆ 서울시 중랑구 면목동
- 용도 ┆ 단독주택
- 대지면적 ┆ 62.8m²
- 건축면적 ┆ 32.7m²
- 연면적 ┆ 93.98m²
- 규모 ┆ 지상 4층
- 높이 ┆ 11.4m
- 주차 ┆ 1대
- 건폐율 ┆ 52.07%
- 용적률 ┆ 149.65%
- 구조 ┆ 철근콘크리트구조
- 외부마감 ┆ 벽돌
- 내부마감 ┆ 석고보드 위 수성페인트
- 구조설계 ┆ 터구조
- 기계·전기설계 ┆ 하나기연
- 시공 ┆ 무원건설
- 설계기간 ┆ 2015. 7. - 2016. 2.
- 시공기간 ┆ 2016. 2. - 8.
- 사진 ┆ 노경
- 건축주 ┆ 김욱겸

어라운드 사옥
- 설계 ┆ 푸하하하프렌즈(윤한진, 한승재, 한양규)
- 위치 ┆ 서울시 마포구 연남동
- 용도 ┆ 단독주택, 근린생활시설
- 대지면적 ┆ 99.5m²
- 건축면적 ┆ 59.66m²
- 연면적 ┆ 246.63m²
- 규모 ┆ 지상 6층, 지하 1층
- 높이 ┆ 17.3m
- 주차 ┆ 2대
- 건폐율 ┆ 59.96%
- 용적률 ┆ 190.27%
- 구조 ┆ 철근콘크리트조
- 외부마감 ┆ 타일
- 내부마감 ┆ 수성페인트
- 구조설계 ┆ 터구조
- 기계·전기설계 ┆ 하나기연
- 시공 ┆ 제이아키브
- 설계기간 ┆ 2016. 1. - 9.
- 시공기간 ┆ 2016. 10. - 2017. 7.
- 사진 ┆ 김용관
- 건축주 ┆ 송원준

GROWN HOUSE
- Architect ┆ FHHH friends (Hanjin Yoon, Seungjae Han, Yangkyu Han)
- Location ┆ Myeonmok-dong, Jungnang-gu, Seoul, Korea
- Programme ┆ housing
- Site area ┆ 62.8m²
- Building area ┆ 32.7m²
- Gross floor area ┆ 93.98m²
- Building scope ┆ 4F
- Height ┆ 11.4m
- Parking capacity ┆ 1
- Building coverage ┆ 52.07%
- Floor area ratio ┆ 149.65%
- Structure ┆ RC
- Exterior finishing ┆ brick
- Interior finishing ┆ painting on gypsum board
- Structural engineer ┆ Teo Structure
- Mechanical and electrical engineer ┆ HANA Consulting Engineers Co., LTD.
- Construction ┆ Moowon Construction Co., LTD.
- Design period ┆ July 2015 – Feb. 2016
- Construction period ┆ Feb. – Aug. 2018
- Photograph ┆ Kyung Roh
- Client ┆ Wookgyeok Kim

AROUND MAGAZINE OFFICE BUILDING
- Architect ┆ FHHH friends (Hanjin Yoon, Seungjae Han, Yangkyu Han)
- Location ┆ Yeonnam-dong, Mapo-gu, Seoul, Korea
- Programme ┆ housing, retail
- Site area ┆ 99.5m²
- Building area ┆ 59.66m²
- Gross floor area ┆ 246.63m²
- Building scope ┆ B1, 6F
- Height ┆ 17.3m
- Parking capacity ┆ 2
- Building coverage ┆ 59.96%
- Floor area ratio ┆ 190.27%
- Structure ┆ RC
- Exterior finishing ┆ tiles
- Interior finishing ┆ painting
- Structural engineer ┆ Teo Structure
- Mechanical and electrical engineer ┆ HANA Consulting Engineers Co., LTD.
- Construction ┆ Jarchiv
- Design period ┆ Jan. – Sep. 2016
- Construction period ┆ Oct. 2016 – July 2017
- Photograph ┆ Yongkwan Kim
- Client ┆ Wonjun Song

콜렉티보 커피 로스터스
- 설계 ┆ 푸하하하프렌즈(윤한진, 한승재, 한양규)
- 위치 ┆ 서울시 용산구 한남동
- 용도 ┆ 제1종 근린생활시설
- 대지면적 ┆ 102.58m²
- 건축면적 ┆ 61.48m²
- 연면적 ┆ 186.24m²
- 규모 ┆ 지상 2층, 지하 1층
- 높이 ┆ 9.5m
- 주차 ┆ 1대
- 건폐율 ┆ 59.93%
- 용적률 ┆ 119.87%
- 구조 ┆ 연와조, 철골기둥
- 외부마감 ┆ 치장벽돌
- 내부마감 ┆ 치장벽돌
- 구조설계 ┆ 터구조
- 기계·전기설계 ┆ 푸하하하프렌즈
- 시공 ┆ 제이아키브
- 설계기간 ┆ 2017. 4. - 9.
- 시공기간 ┆ 2017. 9. - 11.
- 사진 ┆ 노경
- 건축주 ┆ 이승연

성수연방(리노베이션)
- 설계 ┆ 푸하하하프렌즈(윤한진, 한승재, 한양규)
- 위치 ┆ 서울시 성동구 성수이로14길 14
- 용도 ┆ 제2종 근린생활시설, 공장
- 대지면적 ┆ 1,443m²
- 건축면적 ┆ 832.2m²
- 연면적 ┆ 1,994.8m²
- 규모 ┆ 지상 3층
- 높이 ┆ 11.8m
- 주차 ┆ 12대
- 건폐율 ┆ 57.67%
- 용적률 ┆ 138.23%
- 구조 ┆ 철근콘크리트구조, 철골조
- 외부마감 ┆ 노출콘크리트 위 오일스테인
- 내부마감 ┆ 기존 콘크리트 면 정리
- 구조설계 ┆ 센구조
- 기계·전기설계 ┆ 하나기연
- 시공 ┆ 에스엠건설
- 설계기간 ┆ 2017. 9. - 2018. 4.
- 시공기간 ┆ 2018. 5. - 12.
- 사진 ┆ 석준기
- 건축주 ┆ 오티디코퍼레이션

COLECTIVO COFFEE ROASTERS
- Architect ┆ FHHH friends (Hanjin Yoon, Seungjae Han, Yangkyu Han)
- Location ┆ Hannam-dong, Yongsan-gu, Seoul, Korea
- Programme ┆ retail
- Site area ┆ 102.58m²
- Building area ┆ 61.48m²
- Gross floor area ┆ 186.24m²
- Building scope ┆ B1, 2F
- Height ┆ 9.5m
- Parking capacity ┆ 1
- Building coverage ┆ 59.93%
- Floor area ratio ┆ 119.87%
- Structure ┆ brick structure, steel frame
- Exterior finishing ┆ brick
- Interior finishing ┆ brick
- Structural engineer ┆ Teo Structure
- Mechanical and electrical engineer ┆ FHHH friends
- Construction ┆ Jarchiv
- Design period ┆ Apr. - Sep. 2017
- Construction period ┆ Sep. - Nov. 2017
- Photograph ┆ Kyung Roh
- Client ┆ Seungyeon Lee

SEONG-SU-YEON-BANG (RENOVATION)
- Architect ┆ FHHH friends (Hanjin Yoon, Seungjae Han, Yangkyu Han)
- Location ┆ 14, Seoungsui-ro 14gil, Seongdong-gu, Seoul, Korea
- Programme ┆ retail, factory
- Site area ┆ 1,443m²
- Building area ┆ 832.2m²
- Gross floor area ┆ 1,994.8m²
- Building scope ┆ 3F
- Height ┆ 11.8m
- Parking capacity ┆ 12
- Building coverage ┆ 57.67%
- Floor area ratio ┆ 138.23%
- Structure ┆ RC, S
- Exterior finishing ┆ oil stain on exposed concrete
- Interior finishing ┆ clearing the concrete faces
- Structural engineer ┆ SEN Engineering Group
- Mechanical and electrical engineer ┆ HANA Consulting Engineers Co., LTD.
- Construction ┆ SM Construction
- Design period ┆ Sep. 2017 - Apr. 2018
- Construction period ┆ May - Dec. 2018
- Photograph ┆ Joonki Seok
- Client ┆ OTD Corp.

디스이즈네버댓 사옥 (리노베이션)

- 설계 ¦ 푸하하하프렌즈(윤한진, 한승재, 한양규)
- 위치 ¦ 서울시 서대문구 연희로11라길 10-2
- 용도 ¦ 단독주택, 제2종 근린생활시설
- 대지면적 ¦ 298.9m²
- 건축면적 ¦ 154.54m²
- 연면적 ¦ 378.4m²
- 규모 ¦ 지상 3층
- 높이 ¦ 12.1m
- 주차 ¦ 4대
- 건폐율 ¦ 51.7%
- 용적률 ¦ 126.6%
- 구조 ¦ 철근콘크리트구조
- 외부마감 ¦ 접합유리, 베이스 판넬
- 내부마감 ¦ 석고보드 위 수성페인트
- 구조설계 ¦ 터구조
- 기계·전기설계 ¦ 하나기연
- 시공 ¦ 이각건설
- 설계기간 ¦ 2018. 3. - 8.
- 시공기간 ¦ 2018. 9. - 2019. 4.
- 사진 ¦ 신경섭
- 건축주 ¦ 조나단

THISISNEVERTHAT OFFICE BUILDING (RENOVATION)
- Architect ¦ FHHH friends (Hanjin Yoon, Seungjae Han, Yangkyu Han)
- Location ¦ 10-2, Yeonhui-ro 11ra-gil, Seodaemun-gu, Seoul, Korea
- Programme ¦ housing, retail
- Site area ¦ 298.9m²
- Building area ¦ 154.54m²
- Gross floor area ¦ 378.4m²
- Building scope ¦ 3F
- Height ¦ 12.1m
- Parking capacity ¦ 4
- Building coverage ¦ 51.7%
- Floor area ratio ¦ 126.6%
- Structure ¦ RC
- Exterior finishing ¦ glass, base panel
- Interior finishing ¦ painting on gypsum board
- Structural engineer ¦ Teo Structure
- Mechanical and electrical engineer: HANA Consulting Engineers Co., LTD.
- Construction ¦ Leegak Construction
- Design period ¦ Mar. — Aug. 2018
- Construction period ¦ Sep. 2019 — Apr. 2019
- Photograph ¦ Kyungsub Shin
- Client ¦ Jonathan

크리틱

# 두 번 다시 웃지 않는 사나이[1]
: 정지돈



Critique

# The Man Who Never Laughed Again[1]
: Jidon Jung

정지돈

대학에서 영화와 문예창작을 공부했다.
『2015년 문학과 사회 신인문학상』으로
등단, '2015년 젊은작가상 대상' '2016년
문지문학상'을 수상했다. 2018년 베니스
건축 비엔날레 한국관 작가로 참여했다.
지은 책으로는 『내가 싸우듯이』
『우리는 다른 사람들의 기억에서 살
것이다』『작은 겁쟁이 겁쟁이 새로운
파티』『문학의 기쁨』(공저) 등이 있다.

20대 초반, 나는 건축학과에 다니던
친구들인 재경이, 현기, 상민이와 꽤
오랫동안 함께 살았다. 그들은 자신을
예술가로 생각했고 예술대학 학생이었던
내게 여러 의견을 피력했다. 예대 애들은
부정확하고 감상적이다, 게으르고 공부를 안
한다. 그에 반해 건축학과에 다니는 우리는
철저하고 성실하며 심지어 예술적이기까지
하다. 재경이는 사회학자 이진경의
『노마디즘』과 칼럼니스트 김규항의『B급
좌파』를 끼고 다니며 리좀Rhizome적이며
혁명적인 건축에 대해 설파했고 현기는
공모전에 제출하기 위해 빙하 아래에
유리로 만들어진 예술센터를 설계하고는
〈해체/디컨스트럭티브Deconstructive〉라고
이름 붙였다. 상민이는 프랑스 화가 이브
클랭Yves Klein을 흉내 낸 정체 모를 푸른색
물감 더미를 그리고는 유동하는 건축이
바로 비어 있는 건축이라고 말했다. "그게
무슨 말이야?" 내가 물었고 상민이는 대답

Jidon Jung

Jidon Jung studied Film and
Creative Writing. He made his
debut in the professional
literature world as his
short story 『A Blind Owl』
was selected for the New
Writer's Awards of 'Literature
and Society', the quarterly
literary magazine published
by Moonji Publishing Company.
He was awarded the Best of
2015 Young Writers Prize for
『Architecture or Revolution』
and 2016 Moonji Literary
Prize for 『Pale horses』. And
he participated in venice
biennale of architecture 2018
as an artist. He published a
collection of short stories
『Like I Figh』 and a fiction
『A Little Coward Coward New
Party』 and 『We shall survive
in the memory of others』 and
co-author of a literature
critique 『The Pleasure of
Literature』.

In my early twenties, I spent most
of my time with friends Jaekyeong,
Hyunki, and Sangmin from the
architecture department. They
thought of themselves as artists,
condescending to me, an arts school
student, with all sorts of beliefs
about the study of art. They
would say, "Art school students
are inaccurate and sentimental,
they're naive and they don't study.
On the other hand, we are well-
disciplined, dedicated and even
artistic." Jaekyeong would go around
professing the values of rhizomatic
and revolutionary architecture, with
books like 『Nomadism』 or 『Type-B
Gauchiste』 tucked under his arm, and
Hyunki designed a glass arts center
placed underneath an iceberg for a
competition, naming it "Haechae/
Deconstructive." Sangmin would paint
clumps of unidentifiable blue paint
in imitation of Yves Klein, declaring
that moving, flowing architecture is
rightfully empty architecture. "What
does this mean?" I asked, and instead
of answering me, Sangmin quoted
Gaston Bachelard's 『Air and Dreams』.

대신 프랑스 철학자 가스통 바슐라르Gaston Bachelard의 『공기와 꿈 L'air et les Songes』을 펼쳤다.

> "깊은 하늘을 본다는 것은 온갖 인상들 중에서 어떤 감정에 가장 가까운 인상을 준다. … 그것은 감정과 보는 것의 결정적인 융합이며 완벽한 결합이라고 해야 할 것이다."[2]

돌이켜보면 친구들은 예술에 대해 전혀 모르거나 전적으로 오해하고 있었던 것 같다. 친구들 때문인지 나 역시 건축과 건축가에 대해 아무것도 알 수 없었다. 건축은 그럴듯한 외관을 짓고 유행하는 개념을 붙이면 되는 거 아니야? 그게 아니면 그럴듯한 외관을 짓고 참여적인 포즈를 취하거나 그것도 아니면 그럴듯한 외관을 짓고 초연한 척 아래위로 검은 옷을 입고 인터뷰를 피해 다니면 되는 거 아닌가. (물론

인터뷰를 모두 피해선 안 된다. 인터넷에 돌아다닐 잘 나온 사진 몇 장은 남겨야 한다.) 그런 측면에서 예술가와 건축가는 유사한 것 같다. 뭔지 잘 모르겠지만 그럴듯하고 모호하지만 있어 보이는 무엇. 그러나 정말 그런 게 예술이었다면 나는 일찍이 이 모든 것에서 손을 떼었을 것이다. 마찬가지로 그런 게 건축이었다면 나는 건축에 관한 관심을 진작에 잃었을 것이다.

아포리즘 1
무엇이 옳은지는 항상 알 수 있다.
그러나 무엇이 가능한지에 대해서 항상 알 수 있는 것은 아니다.[3]

푸하하하프렌즈를 처음 안 것은 2015년이다. 소설 『건축이냐 혁명이냐』로 운 좋게 상을 받고 주목받기 시작한 나는 문예지에서 비평면을 청탁받았다. 매 계절 나오는 신간 소설의 서평을 써달라는 청탁이었다.

> "The sight of a profound sky is, of all impressions, the closest to a feeling. … It is more a feeling than a visual thing, or, rather, it is the definitive fusion, the complete union of feeling and mediation."[2]

Looking back, these friends knew nothing about art, or else they were fundamentally confused. Perhaps it is because of these friends that I myself do not know anything about architecture and architect. Isn't architecture about building something that looks pleasing, and then furnishing it with some kind of trendy concept? Or is it about building something that looks nice and posing as if you're engaged seriously in something? Or better, build something nice and dress all in black, from top to bottom, to look as aloof as possible, and elude all interviews (naturally, not every interview; you need a few flattering photos online).

From this angle, artists and architects are similar. You don't really know what they are talking about, but it sounds fancy and ambiguous and somehow attractive. Yet, if that were really art, I would long since have left these pursuits. In the same way, if all that architecture was, I would have lost all my curiosity about architecture early on.

APHORISM 1
YOU ALWAYS KNOW WHAT'S RIGHT. YET, YOU CAN NEVER (ALWAYS) KNOW WHAT'S POSSIBLE.[3]

I first met FHHH friends in 2015. Luck had struck, and I had won an award for 『Architecture or Revolution』 and I was starting to get noticed when a literature journal asked me to write a review. They asked that I review a newly published novel, a seasonal affair. A fellow critic and I rifled through the newly minted pages, and that book turned out to be a book of short stories called 『Really Stupid』 by Seungjae Han, one of the members of FHHH friends. The press material read as follows. "An entertaining idea, engaging narrative, an unconvoluted fresh voice/tone, a distinct study

동료 비평가와 함께 신간을 뒤적였고
그때 걸린 책이 푸하하하프렌즈의
세 소장 중 한 사람인 한승재의 소설집
『엄청명충한』이었다. 보도자료에는 이렇게
쓰여 있었다. "흥미로운 착상, 흡입력 있는
전개, 누구의 영향도 받지 않은 신선한 문체,
인간의 본성에 관한 독특한 관찰." 나는
소설집을 읽고 애매한 칭찬이 섞인 비판을
남겼다. "미국 SF 소설가가 쓴 단편소설
같다, 다만 번역이 잘못되었을 뿐." 동료
비평가는 "첫잡네조"라는 평을 남겼다. 록
앨범의 황금률로, 첫 곡으로 귀를 잡아끌고
네 번째 곡으로 조진다는 뜻이라고 한다.
왜 이런 평이 나왔는지 모르겠지만 앨범에
대한 이야기를 꺼냈다는 사실만으로 동료
비평가가 얼마나 옛날 사람인지 알 수 있을
것이다. 문예지에는 한승재가 당시 진행한
인터뷰도 인용되어 있었다. 그는 인터뷰에서
어머니가 등단 소설가인데 등단인가 뭔가가
요즘 시대에 뭐가 필요한지 모르겠다, 소설을

읽었는데 이 정도면 나도 쓰겠다고 생각해서
썼다고 말했다.

아무튼 그 소설집에 대해 평하면서
나는 자연스레 한승재의 본업이 건축가이고
설계사무소 이름이 푸하하하프렌즈라는
사실을 알게 되었다. 대부분 사람이
그렇듯 이름에 대한 첫인상은 좋지 않았다.
트위터에서 봤는데 푸하하하프렌즈에
전화하면 직원들도 회사명을 발음하기
부끄러워 "푸하ㅎ프렌즈입니다."라고
얼버무린다고 한다. 내가 글을 쓴 책은 문학
평론집이었지만 푸하하하프렌즈의 초기 작품
사진도 삽입되어 있다.

이후 한승재와 연이 닿거나 하는
일은 없었다. 가끔 오가는 카페나 건물이
푸하하하프렌즈의 작업이라는 사실을
접하면서 그들이 어딘가에서 잘 살고 있으며
느낌상 점점 더 잘 살고 있는 거 같다는
것 정도만 알았을 뿐이다. 가끔 한승재가
『엄청명충한』 이후 출간하겠다고 한 소설

of the human essence." I read the anthology and wrote a review muddled with awkward compliments. "Feels like a short story written by an American Science fiction writer, yet as if it hasn't been translated properly." My fellow critic left the words "seduce first kill later". This is supposed to be the golden rule of rock albums, where you seduce the listener with the first song, and the final blow comes at the fourth. I have no idea why he decided to write about this, but just from the fact that he decided talk about rock albums reveals how old-fashioned he was. In the review, there is quotes from a recent interview with Han. Han talked about how his mother was a published novelist, and he had wondered what it meant, to be published, and what was required of this generation. This had lead him to read the novel, and, concluding that it was a job easily done, he wrote a novel.

All in all, reviewing that anthology naturally led me to learn that Seungjae Han is originally an architect, and that the name of his firm is FHHH friends[4], like most people, this name did not really give me the best first impression. I saw somewhere on twitter that if you ever call the firm, even the staff feel embarrassed about pronouncing this properly, stumbling over the third and fourth "Ha"s. I wrote the anthological literature review, and some of FHHH friends earlier works were included as photos.

Thereafter, I hardly came across Seungjae Han. Learning that the cafés or buildings I would pass by were projects designed by FHHH friends, I knew only that they were doing well, and going from strength to strength. From time to time, I was curious to learn about the novel 「He's He and He's He」 that he had wanted to publish after 『Really Stupid』 but I never looked it up.

It was around this time that Seungjae Han contacted me. He explained that they had been awarded the Korean Young Architect Award, and that a book would be coming out. Therefore he asked me could you write a review about our works?

I was surprised and rather moved. He was a person with depth, to

「개는 개고 개는 개다」의 소식이 궁금했지만 찾아보지는 않았다.

그러던 중 한승재에게 연락이 왔다. '2019 젊은건축가상'을 수상하게 되었고 그래서 책을 만들 예정인데 혹시 자신들의 작업에 대해 평론을 써줄 수 있겠는가 하는 것이었다.

나는 조금 놀랐고 약간 감동했다. 자신의 소설을 비판한 사람에게 정중한 부탁을 할 수 있는 그릇의 사람이라니. 소설가로서는 상상할 수 없을 정도의 큰 그릇이다. 물론 한승재는 소설가가 아니고 가끔 소설가일 수도 있지만, 아무튼 대부분의 경우 소설가가 아니긴 하지만 그래도 그릇이 크구나. 그러나 그의 사람됨이 청탁을 수락하는데 중요한 요인은 아니었다. 건축에 관한 글을 가끔 쓰긴 했지만 본격적인 평론은 쓰지 않았고 쓸 수도 없다고 생각했다. 책에 실리는 글이 꼭 평론의 성격이 아니어도 좋다고 한승재는 말했지만,

그럼에도 이건 성격이 다른 일이다. 그러므로 나는 판단해야 했다. 이 글을 쓰는 것이 옳은 일인지 그른 일인지가 아니라 가능한지 불가능한지에 대해. 그가 소설을 썼을 때 그것의 옳고 그름을 따지지 않고 가능하기 때문에 쓴 것처럼 내가 푸하하하프렌즈와 그들의 건축에 대한 글을 쓰는 것이 가능한 일인지에 대해.

아포리즘 2
오로지 미래의 관점에서만 우리는 비판적 간격을 재획득할 수 있다.[4]

그래서 어쩌면 푸하하하프렌즈에 대해 말하기 위해 건축을 새롭게 발명해야 하는지도 모른다. 건축이라는 분야 자체를 새롭게 혁신한다는 의미에서가 아니라 우리가 흔히 말하는 건축, 전문가들 사이에서 말하는 건축이나 대중들이 말하는 건축에서의 건축이 아닌 다른 무언가를

make this request to a person who had criticized his novel. For a novelist, such depth is unimaginable. Of course, Seungjae Han is no novelist, and while he might sometimes be a novelist, most of the time he isn't. Still, he's a person with depth. It wasn't his humility that moved me to accept his request. While I had dabbled in writing about architecture, I had never actually critiqued architecture, and I also believed that I couldn't. Han told me that there was no need for the article to fit the format of a critique, but I felt that it was a task of an entirely different nature. I had to make a decision. It wasn't so much a question of whether writing would be right or wrong, but rather whether it would be possible or impossible. Just as I had written a review about his novel, not because it was right or wrong to do so, but because I could, I had to decide whether it would be possible for me to write about FHHH friends and their architecture.

APHORISM 2
IT IS ONLY FROM BEING SITUATED IN THE FUTURE, THAT WE CAN RE-ACQUIRE THE DISTANCE REQUIRED TO CRITIQUE.[5]

Perhaps, then, for me to talk about FHHH friends, architecture must be re-invented. This does not mean that the genre itself must be transformed anew, but that a certain something must be established; a certain something which deviates from the architecture we talk about, the architecture discussed among architects, and the architecture discussed by the masses.

This also applies to literature. Before, deugndan[6] was a necessary process. You were officially recognized as an author, and then through the national annual spring literary contest, you published with a major publisher, and you were critiqued by a critic with a PhD, and you got a famous literature award⋯⋯ This is all still meaningful, yet the field of literature has seen a fundamental change in the way they think. That feeling we all get when we read and write is, at the end of the day, not about all that. The field has belatedly realized that the

만들어내야 한다는 의미에서.

　　문학도 마찬가지다. 예전에는 등단이
필수 절차였다. 주요 일간지의 새해 문예
당선자로 뽑혀 '신춘문예'로 등단하고 대형
출판사에서 책을 내고 박사 출신 비평가에게
평론을 받고 유명 문학상을 받고…… 지금도
그런 게 의미가 있지만, 문학계에는 뭔가
근본적인 인식의 변화가 일어났다. 우리가
책을 읽고 쓰는 것에서 오는 좋음은 그런
문제가 아니지 않나. 복합적인 좋음과 싫음,
감정과 앎이 어떤 정규 과정이나 관습적
세계 속에 녹아 있었다는 사실을 뒤늦게
깨달은 것이다. 그러므로 문학을 재발명하지
않으면 당대의 작품에서 우리가 느끼는
무언가를 제대로 이야기할 수 없다.

　　그런 의미에서 푸하하하프렌즈는
건축적 재발명이 필요한 형상이다. 그들은
건축가 김수근의 제자도 아니고 김수근의
제자의 제자도 아니고 네덜란드 건축가
렘 콜하스Rem Koolhaas의 제자도 아니고

기존의 장場에서 봤을 때 어딘가 이질적인
존재들이니까. 그런데 정말 그럴까? 단지
'이름' 때문에 그렇게 느껴지는 건 아닐까?
혹은 이질적이라는 건 대체 뭘까? 사람들은
이질적인 것을 원하는 걸까, 꺼리는 걸까?
아니면 둘 다일까?

아포리즘 3
사상은 우리를 나누고 꿈은 우리를 합친다.[5]

너무 거창해진 것 같다. 뭘 건축을
발명씩이나…… 그러나 아주 작은 것, 사소한
것도 모두 발명의 산물이라 생각한다면
이것은 자연스러운 일일지 모른다. 그리고
푸하하하프렌즈는 거창하면서도 사소한
사람들이다. 최소한 첫인상은 그랬다.

　　그들은 우선 컸다. 키가 큰 건지 몸이
큰 건지 얼굴이 큰 건지 알 수 없었지만 커
보였고 행동은 거침없었다. 처음 그들의 설계
사무실에서 만났을 때는 일종의 역할극이나

comprehensive like and dislike, what
is felt and what is known was all
amalgamated into some sort of official
process or ritualistic world. Hence,
without reinventing literature, it is
impossible to discuss that which we
feel from the greatest works of our
age.

　　It is in this sense that, FHHH
friends seems to need an architectural
reinvention. These architects are
not the disciples of Swoogeun Kim,
or the disciples of the disciples
of Swoogeun Kim. They are not the
disciples of Rem Koolhaas, and they
are, in fact somewhat contrarian.
But, is this really true, or is it
just the way the name makes us feel?
What does it mean to be heterogenous
and do people want the heterogenous?
Do they dislike it, or is it both?

APHORISM 3
IMAGINATION DIVIDES US AND DREAMS
UNITE US.[7]

Perhaps that's going a little too
far. Reinventing architecture does
seem a little excessive…… Yet, if we
think that the smallest of things,

the trivial, are all the living
products of invention, this is a
natural outcome. And FHHH friends
are excessive yet trivial people. At
least, that is the first impression
that they make.

　　First, they were big. It's hard
to say if this impression comes from
their height, or because of being
well built or even if it's just
that their faces were big, but they
looked big to me, and their behavior
was without hesitation. When I first
met them at the firm, it felt like
I was watching some sort of play
or entertaining talk. Seungjae Han
was kindly, cracking his cute but
decidedly unfunny jokes, and Yangkyu
Han complained about Seungjae Han
with a cynical expression, while
Hanjin Yoon just did his own thing.
The three were very different from
each other, but also so well suited,
and watching them, I visualized
shapes, and these shapes were
similar to what I had felt when I
had received the buildings that each
of them had designed. Before I get
into the buildings and shapes, let me
quickly tell you about how they work.

만담을 보는 듯했다. 한승재는 친절한 태도로 귀엽지만 안 웃긴 농담을 계속했고 한양규는 시니컬한 얼굴로 한승재를 구박했으며 윤한진은 자기 갈 길을 갔다. 세 사람은 몹시 달랐는데 잘 어울렸고 나는 그들을 보며 도형을 떠올렸다. 그것은 그들 각자가 설계한 건물을 봤을 때 느낀 특성과 유사했다.

　　도형과 건물을 말하기 전에 그들의 작업 방식에 대해 간단히 이야기해야겠다. 그들은 프로젝트가 들어오면 해당 프로젝트에 어울릴 것 같은 사람을 정한다. 특별히 협업하지는 않으며 서로의 작업을 보고 비난하거나(비판 말고) 비웃거나 멸시하는 등의 과정을 거친다(또는 견딘다). 이 과정 때문에 푸하하하프렌즈의 구성원에게는 정신력이 중요하다. 물론 몇 년 동안 함께 작업하며 이제는 서로를 존중하는 법을 배웠지만 말이다. 나는 동료이자 친구인 이들이 초기와 달리 서로의 작업을 존중하는 것은 그들이 인격적으로 성숙해졌기 때문만이 아니라 작업의 질 자체가 성장했기 때문이 아닐까 생각했다. 하지만 이런 말을 건네지는 않았다. 그들은 칭찬을 어색해하는 사람들이었고 혹은 칭찬하면 지나치게 있는 그대로 받아들이는 사람들이었다. 아무튼 다시 돌아가 그들에게는 각자 어울리는 도형이 있다. 윤한진은 세모, 한승재는 원, 한양규는 네모.

　　나와 그들은 2019년 8월의 어느 여름날 만나 연남동과 연희동을 돌아다니며 건물들을 봤다. 윤한진이 설계한 〈어라운드 사옥〉(세모), 한양규가 설계한 〈디스이즈네버댓 사옥〉(네모), 한승재가 설계한 〈연희동 주택〉(원).

　　〈연희동 주택〉은 아직 공사 중이었지만 한승재의 애정은 대단해보였다. 주택은 그의 본가에서 가까운 곳에 있었고 그는 누구보다 연희동과 그곳 계단과 골목에 익숙한 듯 자연스럽게 돌아다니며 동네와 건물을 설명했다. 주택은 바깥과 단절되어

When a project arrives at the firm, they decide who suits it best. They don't really work together, and each go through (or tolerate) a process of having their work criticized (not critiqued) and smirked at and hated on. Due to this process, mental strength is an important element of being a member of FHHH friends. Of course, while working together for the past couple of years, they have now learnt how to honor each other. I thought to myself that perhaps these guys, as colleagues and friends, had come to honor each other's work, unlike during their early years, not because they had built strength of character, but because the quality of their work itself had grown. I did not mention this to them. They were the type to feel uncomfortable around compliments, and, if complimented, had the tendency to overwhelmingly take it as given.

　　Anyway, returning to the question at hand, each person has a distinctive shape. Hanjin Yoon: a triangle, Seungjae Han: a circle, Yangkyu Han: a square.

　　I met with them on a summer day in August, and we went to see buildings in and around Yeonnam and Yeonhee-dong. <Around Magazine Office Building> (a triangle) designed by Hanjin Yoon, <Thisisnverthat Office Building> (a square) designed by Yanggyu Han, and <Yeonhee-dong House> (a circle) designed by Seungjae Han.

　　The <Yeonhee-dong House> was still under construction, but Seungjae Han's affection seemed great. The house was located close to Seungjae Han's house, and he described the area of Yeonhee-dong and buildings as if he were familiar with the stairs and alleys of the space than anyone else. The house was a space disconnected from the outside, but it felt like the outside landscape had penetrated the inside, embraced the outside, and felt like a stairway from the outside and an alley in the alley. We walked inside and out, eating the rice cake handed out by the construction workers. <Thisisneverthat Office Building> was solid, clean and cool. It was hot, but it was as pleasant as entering the valley when I entered the building entrance. I wanted to open the made-in-Germany fridge which

있었지만 바깥 풍경이 안으로 스며들어온 느낌이었고 안에서 밖을 품은 느낌이었으며 밖에서 안으로 난 계단, 골목 안의 골목처럼 느껴졌다. 우리는 현장 작업자들이 건네준 떡을 먹으며 안과 밖을 거닐었다.

〈디스이즈네버댓 사옥〉은 단단하고 깨끗하고 시원했다. 더운 날씨였지만 건물 입구로 들어서자 계곡에 들어선 것처럼 쾌적했다. 연희동에 갑자기 등장한 독일제 냉장고(?)의 냉동실 문을 열고 고개를 넣은 채 한동안 가만히 머물고 싶었고 옥상으로 가는 계단에서 보이는 하늘은 협곡 사이에서 올려다본 풍경처럼 서늘하고 고요했다.

〈어라운드 사옥〉은 자신만만했다. 보는 위치에 따라 날카롭거나 위태롭고 소박하거나 모던하게 인상이 바뀌었지만, 어떤 관점에서도 자신의 자세에 날이 선 자신감을 유지하는 모습이었다. 내부는 외관과 달리 귀여웠고 우리는 어라운드의 직원들과 고양이 가족을 피해 다니며 건물

곳곳을 만지고 바라보았다.

나는 한승재, 한양규, 윤한진과 칼국수를 먹었고 중간에 잠깐 들린 한승재의 집에서 수박을 먹었으며 한승재 조카가 그린 그림을 보고 그의 어머니와 아버지를 뵙고(?) 어머니가 1998년에 출간한 수필집 『게으름 피우기』에 사인도 받았다(??). 너무 순식간에 여러 일이 일어나 정신이 없었지만 그들은 익숙해보였다. 많은 이야기를 들었지만 기억에 남는 건 대부분 그들이 질색하는 것들에 관한 이야기였다. 푸하하하프렌즈는 좋아하는 건 겹치지 않지만 질색하는 건 겹치는 사람들이다. 공통의 적이 있는 것이야말로 뭉치기에 최적의 요건이다. 그런 의미에서 한국은 최적의 장소다. 거리에 나가면 온통 적들뿐이니까. 그러므로 승재와 양규와 한진의 꿈은 같다. 적들을 무찌르고 질색하는 것을 더는 보지 않기.

suddenly appeared in the middle of Yeonhee-dong, to put my head in the freezer and stay like that for a sustained period of time. The sky I saw on the way up to the rooftop was shadowy and serene like looking up at the sky from between two ravines. The atmosphere at <Around Magazine Office Building> was brimming with confidence. Its impression would change according to your location, as acute or risky or modest or modern, but its very stature seemed to own an acute sense of confidence. The interior, unlike the exterior, was cute, and we touched and looked at a variety of different places in the building while trying to avoid the staff and a family of cats.

I ate *kalguksu* (noodles) with Seungjae Han, Yangkyu Han and Hanjin Yoon, and during a fly-by visit to Seungjae Han's house, we ate watermelon and looked at the pictures drawn by a nephew, and met his mother and father, and I also received an autographed copy his mother's 1998 essay collection 『Being lazy』. All of this happened so fast that for me, it felt quite hectic, but it seemed like

they were used to this. I listened to a lot of stories, but the most memorable are the stories I heard about their deep hate about certain things. FHHH friends do not have mutual likes, but they are united in their mutual hates. Indeed, one of the most optimal conditions to band together is to have a common enemy. In that sense, Korea is an optimal place. Out on the streets, everything is an enemy. Hence, this is like the dream of Seungjae and Yangkyu and Hanjin. Defeat the enemy and never see what they hate again.

APHORISM 4
TO BE AN ARCHITECT, ONE MUST FIRST
BECOME AN OPTIMIST.[8]

"Here is a story that architects like to tell. After a request to work on the extension by a couple, an architect visits them to share a meal. He listens carefully to what they need, and what kind of requirements they have, and the opinions of the husband and wife. After the meal was over, his gives his expert

아포리즘 4
건축가가 되려면
낙관주의자가 되어야 한다.[6]

"한 건축가가 가끔 하던 이야기다.
어느 부부로부터 집을 증축하고
싶다는 의뢰를 받고 찾아간 그는
그들과 함께 저녁 식사를 했다. 그들이
필요한 것이 뭔지, 그리고 어떤 요구
사항이 있는지 잘 듣고, 남편과 아내의
의견도 각각 들었다. 저녁 식사가
끝나고 그는 전문가다운 조언을
제시했다. "두 분에게 증축은 필요
없습니다. 그냥 이혼하시면 됩니다.""[7]

건물의 건축적 의의는 건축 평론가나
기자들이 잘 알 것이다. 실력에 대해서는
건축가가 잘 알 것이며 공간이 실제로
일하거나 살기 어떤지에 대해서는 건축주나
사용자가 잘 알 것이다. 보기 어떤지에
대해서는 인스타그래머가 잘 알 것이고…….
가끔 이것들은 서로 반대 방향을 가리킨다.
건축주는 불편을 호소하지만, 건축가와
평론가는 엄지손가락을 치켜든다. 독일
건축가 미스 반데어로에 Mies van der Rohe의
판즈워스 하우스 Farnsworth House, 프랑스
건축가 르 코르뷔지에 Le Corbusier의 빌라
사보아 Villa Savoye, 미국 건축가 프랭크
로이드 라이트 Frank Lloyd Wright의 여러
주택은 건축주의 희생을 요구했다. 사람들이
몰려들고 인스타그램의 명소가 되지만
건축가는 질색한다. 사용자는 편하지만
건축가는 건물을 다 망쳐놓았다고 푸념하고,
평론가는 감탄하지만 사람들은 이해하지
못한다. 건물은 누구의 관점에서 보아야
이해되는 것일까? 소설은 독자의 관점에서
보면 된다. 평론가의 관점은 또 다른 독자의
관점 가운데 하나일 뿐이다. 소설의 주인은
소설가도 아니고 평론가도 아니고 소설을
읽는 사람일 뿐이니까. 그러면 건물은

---

opinion, 'The two of you don't
need an extension. You need a
divorce.'"[9]

The architectural significance of
a building is more than notoriety
among architectural critics or
journalists. Aptitude can easily be
understood by an architect, and what
it is like to actually work or live
in the spaces that are understood by
the owner or the user. How it looks,
is the domain of the instagrammer…….
Sometimes, these elements contradict
each other. The owner might contest
that its uncomfortable, yet the
architect and the critic might put
their thumbs up. Farnsworth House by
Mies van der Rohe, Villa Savoy by Le
Corbusier, and the many residential
projects of Frank Lloyd Wright all
required sacrifice on the part of the
owner. A project might be visited by
crowds of people, become a select
venue on Instagram, but the architect
detests this. The user may make
themselves comfortable, only for the
architect to complain that the user
has utterly destroyed the building.
The critic might applaud a project
which people don't understand. Through
whose perspective must we look to
understand a building? Fiction
requires only that we look at the
reader's perspective. The critic's
opinion just becomes another reader's
opinion. The owners of a novel are
not the novelists or the critics,
it is the people who read the book.
Then, could we say that the owners of
a building are the people who use it,
and hence it is only their opinion
which judges the right or wrong of a
building? Or, should it be the client
who paid for it? Some buildings look
like they exist entirely to satisfy
the architect. Whose desires do
FHHH's projects fulfill? Do each of
these perspectives exist separately,
or detach, or overlap or coexist?
To whom do those buildings that I
observed and experienced ultimately
belong to?

APHORISM 5
WHICHEVER OAK TREE IT MIGHT BE, ONE
CANNOT PREDICT ITS FINAL FORM.[10]

So, for me, that had always been a
puzzle. The fact that somebody who

건물을 사용하는 사람이 주인이니까 오직 그들의 관점에서 옳고 그름이 판명되는 걸까? 아니면 돈을 들인 건축주의 관점에서? 어떤 건물은 오직 건축가를 만족하기 위해 존재하는 것처럼 보이기도 한다. 푸하하하프렌즈의 작업은 누구의 관점에서 누구를 만족시키고 있는 것일까? 각자의 관점은 따로 떨어져 존재하는 것일까? 중첩되거나 혼재되는 것일까? 내가 잠시 보고 느낀 그들의 건물은 누구의 관점에 속하게 되는 것일까?

아포리즘 5
어느 떡갈나무든 최종적인 형태는
미리 예측할 수 없다.[8]

그러니까 나는 그것이 늘 이상했다. 기념을 위한 건물이라면 상관없지만 주택이나 오피스 빌딩 같은 건물을 사용해보지도 않은 사람이 말한다는 사실이, 그것이 단지 우리

눈에 보인다는 이유로. 공간이 중요하다고 하지만 외관, 형태, 구조에 대한 이야기 말고 어떤 이야기를 할 수 있을까? 도시는 단지 커다란 미술관에 불과할까? 나는 이미 스페인의 구겐하임빌바오뮤지엄 Guggenheim Bilbao Museum 때 있었던 철 지난 문제를 반복하는 것에 불과할까?

푸하하하프렌즈와 대화를 나누고 그들의 건물을 보며 든 생각은 그들이 이런 문제에 대해 누구보다 깊게 생각하고 있다는 점이었다. 우리는 장난기와 진지함을 나란히 두지 못하고 가벼움과 무거움이 공존할 수 없다고 생각한다. 푸하하하프렌즈의 이름이나 그들이 자신을 소개하는 방식은 그들에게 상투적인 수식어들, '기발하고 재치 넘치는' '핫플' '유행에 민감한' '특이한' 등이 붙게 만든다. 그러나 조금만 더 들어가 보면 '이들이 지나칠 정도로 진지하고 그래서 이상해졌구나.' 또는 '이질적으로 보이는구나.' 하는 사실을 알게 된다. 한승재와

hadn't even used a building, would talk about it, just because it could be seen. Even structures that were not monuments, but houses or office buildings. While they say that space is important, what could they talk about other than all that talk about the exterior, the form and the structure? Is the city just a large art museum? Was I simply reliving the Bilbao effect?

Through conversation with FHHH friends and observation of their projects, I have reached the conviction that they have thought more profoundly about this issue than anyone else. We believe that the mischievous and the serious cannot exist side by side, that lightness and weight cannot coexist. The name FHHH friends, or the way they introduce themselves make it easy to add tired adjectives to describe them, like "novel and witty," "hot," "trendy," or "unique." Yet, dig a little deeper, and you come to recognize that they are in fact, too serious, and have become a bit weird. The word that Seungjae Han and Yangkyu Han, Hajin Yoon mentioned

the most was "basics". Actually, Yangkyu Han even quoted John Ruskin's 『The Seven Lamps of Architecture』. This made me nervous that this would lead to Vitruvius' 『The Ten Books on Architecture』. We need to get the basics right, but once we start trying to get the basics right, the basics defined in society were quite far from that good feeling we get when we deal with buildings. What should be considered as fundamental, and how can these be safeguarded? Perhaps that's why they stacked those bricks with their bare hands, and wrote articles and criticized each other and fought and drew pictures and made clothes (Actually, I'm not quite sure why they made the clothes) and hated on each other, and made a list of everything they really hate, and exterminated them one by one, while trying to find their way. And what is left is something that is slightly eccentric, when we look at it, an attitude that can seem like childishness according to how you look at it.

After seeing the Crystal Palace at the London World Expo in 1851,

한양규, 윤한진이 가장 많이 말한 단어는
'기본'이었다. 심지어 한양규는 몇 번이나
문학 평론가 존 러스킨John Ruskin의 『건축의
일곱 등불The Seven Lamps of Architecture』을
인용했다. 나는 이러다 그가 로마시대 건축가
비트루비우스Vitruvius의 『건축십서De
Architectura』까지 인용하는 건 아닐까
긴장했다. 기본을 지켜야 한다. 그런데
기본을 지키려고 보니 사회에서 말하는
기본이 우리가 건물을 대할 때 느낄 수
있는 좋음과 멀리 있었다. 그렇다면 진짜
기본은 무엇일까? 기본은 어떻게 지킬 수
있는 것일까? 그래서 그들은 손으로 벽돌을
쌓아 올리고 글을 쓰고 비난하고 싸우고
그림을 그리고 옷을 만들고(사실 옷은 왜
만들었는지 모르겠지만) 싫어하고 질색하는
수많은 것을 나열하고 그것들을 소거해가면서
길을 찾아갔던 것 아닐까. 그리고 남은 것이
우리가 보았을 때는 조금 별난, 보기에 따라
치기 같기도 한 태도 아닐까.

1851년 영국 런던 만국박람회의 수정궁을
보고 난 뒤 17세의 윌리엄 모리스William
Morris는 이렇게 말했다.

"생산물의 미적 수준은 가증스러울
정도였고 어떤 관점에서 보더라도
잘못되었으며 참담함이 넘쳐나고 너무
지나치고, 너무 과장되었으며……"[9]

푸하하하프렌즈의 질색은 17세의
윌리엄 모리스가 가졌던 분노와 같다.
푸하하하프렌즈의 이름이 어리고
천진해보인다면 그것은 옳다. 다만
그것은 단지 낙관과 즐거움만을 가진
'어림'이 아니라 부당한 무언가에 대한
분노를 가진 '순수함'이다. 왜 저런 건물이
지어지는 것일까? 흉내 내기에 불과한
치장, 현학적이고 자만심만 가득한 외양.
이런 것들은 건축에 대한 잘못된 이해에서
오는 것 아닐까. 건축이란 건축가를 위한

William Morris reportedly said,

> "The quality was aesthetically
> loathsome, wrong from every
> angle, overflowing with
> wretchedness and exaggerated in
> the extreme."[11]

The intense hate of FHHH friends is
similar to the rage which infiltrated
William Morris at the age of 17. If
the name FHHH friends comes across
as young and innocent, that is
true. Yet, this youth is not that
of giddy optimism and joy, but the
innocent passion of anger against
the unjust. Why would a building
like that get built? Decoration that
does little more than imitate, and
exteriors replete with pedantry and
arrogance: Do these come from a wrong
understanding of architecture? If
architecture does not function for
the architects, or its users, or its
critics, then perhaps it is a type
of fallacy resulting from easily
fixating or adapting to one of them.
Hence, at the core of FHHH friends
is their sense of inquisition. That
which must be unceasingly proposed

to find the intersection between the
three sections of architect, user and
critic. That which must be based on
re-inquiring and re-inventing each of
these perspectives. In this sense,
Hanjin Yoon, Seungjae Han, Yangkyu
Han all seem to be in that optimal
place. There are three of them, and
all three are architects, and users
and critics, and can continue to
construct and question, hate and
reconstruct.

것도 아니고 사용자를 위한 것도 아니고
비평가를 위한 것도 아닌데, 어느 것 하나에
쉽게 고정하거나 맞췄을 때 오는 일종의
거짓말 아닐까. 그러므로 푸하하하프렌즈의
기본이란 의문 같은 것이다. 건축가, 사용자,
비평가 세 개의 항이 이루는 접점을 찾기
위해 끊임없이 제기되어야 하는 것. 서로의
관점에서 다시 묻고 재발명해야 하는 것.
그런 면에서 한승재, 한양규, 윤한진은
최적의 위치에 있는 듯하다. 그들은
세 명이고 그들 각자가 건축가이면서
사용자이고 비평가이기 때문에 끊임없이
짓고 질문하고 질색하고 다시 지을 수 있는
것이다.

# 아이디알

# IDR

전보림, 이승환은 서울대학교에서 건축을 공부하고
각각 건축사사무소 M.A.R.U.와 아뜰리에17에서 실무를
익혔다. 2009년 런던으로 이주하여 런던 메트로폴리탄
대학교에서 MA Master of Arts 과정을 마치고 2014년
귀국해 아이디알 건축사사무소를 개소했다. 2017년
첫 준공작인 ‹매곡도서관›으로 ‘신진건축사대상 대상’
‘건축문화대상 우수상’ 등을 수상했다. 각각 서울대학교와
한국예술종합학교에서 설계 스튜디오와 디지털
텍토닉 수업을 담당하고 있으며, 서울시 공공건축가와
행복도시공공건축가로 활동하고 있다. 사용자와 일상을
매개하는 배경으로서 건축의 역할에 관심을 가지고 있으며
블로그를 통해 공공 건축과 건축 설계 현실에 대한
글쓰기를 이어가고 있다.

Borim Jun and Seunghwan Lee studied architecture
at Seoul National University and started their
training at M.A.R.U. and Atelier 17, respectively.
After moving to London in 2009 and receiving
MA degrees at London Metropolitan University,
they returned to Korea in 2014 and founded IDR
Architects. They received several awards for
‹Maegok Library› in 2017, which include the Korean
Rising Architect Award and Korean Architecture
Awards. They have been teaching at design studios
at Seoul National University and Korea National
University of Arts. They are interested in the role
of architecture as a background which mediates
users and their everyday life, and has been writing
about public architecture and the struggles of
young architects on their blog.

# 불만

# Dissatisfaction

Essay One

#질문 #매곡도서관 #다목적강당 #한강건축상상전 #보행교

ㄱ 대규모 재개발을 하면서 좁은
골목길이었던 피맛골을 건물 안의
통로로 바꿔버린 모습. 역사와 문화를
존중하지 않는 개발은 언제나 불만의
대상이다.

언젠가부터 전보림은 마음에 들지 않는 풍경이 눈에 띌
때마다 사진을 찍는 것이 습관이 되었다. 한참 공사 중인
택지 개발 지구의 나무를 밀어버린 황량한 풍경이나 차도만
넓어서 걸어 다닐 기분이 나지 않는 가로 같은 것들이다.
비상식적인 용적률을 자랑하며 병풍처럼 늘어선 아파트도
단골 소재다. 사실 전보림은 얼마 전부터 책을 준비하고 있다.
시기도 출판사도 정해진 건 없지만, 우리네 도시 환경에 관한
불만과 개선안을 담은 책이다. 오가며 찍는 사진은 그 책에
쓸 자료들이다.

　　　전보림은 사회에 불만이 많다. 이승환도 크게 다르지
않다. 가족과 몇몇 지인은 긍정적 시각을 가져보라고
충고하기도 한다. 분명 이런 사회에 대한 불만이 기성세대를
향한 비난으로 느껴질 수도 있다. 하지만 우리는 건축가란
본질적으로 사회에 불만이 많은 사람이어야 한다고 믿는다.
즉 삶과 환경을 결정하는 수많은 조건에 끊임없이 질문을
던지며 더 나아질 수 있는 방법을 찾고 제안하는 직업이라고
생각한다. 대안 있는 불만은 건축가에게 프로젝트를
이끌어가는 원동력이 된다. 새로운 것을 덧붙이기 전에

#Question #MaegokLibrary #MultipurposeAuditorium
#TheHanRiverImaginationExhibition #Footbridge

Borim Jun habitually takes photos whenever she
spots something unappealing. This includes the
desolate image of trees being bulldozed in
residential development districts, or unattractive
streets with a disproportionate amount of space for
car traffic than pedestrians. Another common theme
is apartments, amassed in single file, boasting
irrational FAR ratios. In fact, Borim Jun's book
has now been a while in the making. Despite no set
date or publisher, it's a book aiming to discuss
our dissatisfaction about our cities, and how we
can improve them. The photographs she captures in
the process are all future material for this book.
　　　Borim Jun is extremely dissatisfied with
society. Seunghwan Lee takes a similar stance.
Friends and family have advised both to try being
more optimistic. Of course, such an outward
expression of dissatisfaction about society might
come across as resentment to older, established
generations. However, we believe that all
architects must fundamentally feel dissatisfied
about the state of society. Hence, we believe
the role of the architect is to seek out and
propose ways to improve society by questioning the
countless conditions which determine our lives and
environments. Dissatisfaction, backed up with an
alternative solution becomes the driving motivation

잘못된 것을 고치는 것만으로도 세상에 할 일이 넘치지 않는가.

　우리의 불만은 이미 지금의 사회와 도시에서 당연시하고 있는, 하지만 우리의 가치관으로는 동의할 수 없는 것들에서 비롯된다. 왜 우리가 사는 도시는 차도가 보도보다 넓은가? 왜 공공 건축 대부분에 전면 광장이 있어야 하는 것일까? 로비를 꼭 이렇게 필요 이상으로 넓고 화려하게 만들어야 하나? 도시 경관 차원의 조화를 위해 건물을 더 강하게 규제할 필요는 없을까? 상업 시설에 거대한 간판을 덕지덕지 두르는 것 말고 정말 다른 방법은 없는 것일까? 우리는 이러한 불만을 프로젝트의 구체적 실행 단계에서 전략적으로 활용한다. 때로는 해묵은 불만이 프로젝트를 선택하는 동기가 되고, 건축적 제안과 실행 단계에서 핵심 아이디어가 되기도 한다. 어떻게 보면 불만은 우리가 설계를 시작하는 소중한 첫 단추인 셈이다.

　우리는 아이를 낳아 기르면서 도서관에 자주 다니기 시작했다. 책을 좋아하는 우리와 아이들이 함께 시간을 보낼 곳이 필요했기 때문이다. 그런데 아동 도서와 일반 도서의 서가가 분리되어 있어 각자 보고 싶은 책을 보거나 빌리기 위해 두 개의 열람실을 오가는 것이 영 불편했다.

ㄴ 도시에서 가장 크게 느끼는 불만은 자동차 위주로 개발된 구조이다. 넓은 도로는 도시를 쪼개고 보행자를 위협하며 환경을 망치는 주범이다.

for an architect to lead a project. The world is already overflowing with things that need fixing, even before we attempt to add anything new.

　We live in societies and cities full of seemingly ordinary things that fill us with doubt and challenge our values. Why are the traffic lanes in our cities wider than the pedestrian sidewalks? Why are most public architecture projects attached to a plaza in front? Do lobbies really require such excessively large and luxurious designs? Shouldn't buildings be subject to stricter regulation for the sake of harmonious urban landscapes? Isn't there really any other way than encasing commercial buildings with signs? Our firm strategically introduces these complaints in the day to day elaboration of projects. At times, old battered complaints might emerge as a reason for deciding to work on a project, while it can also become the key concept of the proposition or execution of an architectural project. In a way, dissatisfaction becomes the small yet integral step to launch the design process.

　Ever since we became parents, we regularly visit the library. As book lovers, we needed a place to spend time with our children. Yet, in a library with separate adult and children's sections, moving between the two reading rooms to get to the books we wanted turned out to be quite inconvenient. Also, the library is almost always full of people studying for their exams,

ㄱ The architects' biggest complaint in the city is the urban structure developed mainly for automobiles. Wide roads are dividing the city, threatening pedestrians, and spoils a safe environment.

애당초 도서관에는 책 읽는 사람보다 수험생으로 가득해 늘 정숙을 요구당하는 분위기도 마음에 들지 않았다. 그러던 차에 2014년 설계 사무소를 열고 첫 프로젝트로 도서관 설계 공모가 눈에 들어왔다. 서울에서 먼 울산이었지만 지식에 대한 열린 가능성을 가진 도서관의 매력에 이끌렸다. 그리고 처음 떠오른 생각이 앞서 밝힌 기존 도서관에 대한 불만에서 시작한, 바로 아이와 어른이 함께 책을 읽을 수 있는 도서관이었다. 두 열람실이 막힘없이 하나의 커다란 공간으로 연결되고, 두런두런 책 읽어주는 소리가 백색 소음처럼 바탕에 깔리는 도서관. 그것이 울산시 북구에 지은 ‹매곡도서관›의 첫 아이디어였다.

　　무엇보다 기분 좋은 열람실이 있는 도서관을 만들고 싶었다. 몇 해 전 2010년에 개관한 경기도 판교도서관을 구경한 적이 있었다. 마당에서 지붕까지 걸어 올라갈 수 있고, 외부 공간이 꽤 그럴듯해 보이는 도서관이었다. 입구에 들어서면 머리 위에서 빛이 한가득 떨어지는 아트리움이 방문객을 맞이한다. 그러나 그 멋진 공간은 책 읽는 공간이 아니라 그저 입구와 열람실을 오가는 로비일 뿐이다. 도서관의 주인공은 책 읽는 공간인데, 열람실은 답답하고 특징이 없고 통로 격인 로비가 훨씬 더 화려했다.

rather than people reading books, and we also didn't like feeling pressured to keep quiet. It was at this moment, in 2014, just after we opened our architectural firm, that a library design competition caught our eye as the first project. We were attracted by the idea of designing a library that can enlarge our boundaries of knowledge. And our first inspiration came from our dissatisfaction with existing libraries, inspiring us to imagine a library in which adults and children can read together. The two reading rooms are connected as a large single space with nothing in between, a library where the sound of a parent reading a book to a child hums in the background like white noise. This was our point of departure for the ‹Maegok Library› which we built in Buk-gu, Ulsan.

　　Above all, we wanted to design a library that had a feel-good reading room. A couple of years ago, we visited the freshly built Pangyo Library in Gyunggi-do. It was a library where one could walk from the garden up to the roof, with an extensive and attractive outdoor spaces. Upon entering the site, an atrium illuminated by streams of light from above welcomes visitors. However, this magnificent space was not a space to read books, but a passage used to go from the entrance to the reading room. The main feature of a library should be the reading space, yet, the readings rooms were stuffy and ordinary, and the lobby, a place to simply pass through, was a lot more exquisite.

↗ It is the sketch of the reading room at <Maegok Library> submitted to the competition. What the architects wanted to achieve in <Maegok Library> was to create a library where children and adults could read books together.

매곡도서관 (2017)
열람실은 이 도서관의 중심 공간으로 가장 풍성하고
기분 좋은 분위기를 전한다.

MAEGOK LIBRARY (2017)
The reading room is the central space of the
library and carries the most valuable and
pleasant atmosphere.

‹매곡도서관›은 그다지 크지 않은 구립 도서관이라 모든
공간을 풍요롭게 만들 여유가 없었다. 공공건축 대부분이
그렇듯 선택과 집중이 필요한 프로젝트였다. 그래서 모든
노력과 자원을 도서관의 본질인 책 읽는 공간에 집중했다.
어떻게 보면 너무나 당연한 이 선택이 ‹매곡도서관›을 만든
아이디어의 한 축이라 할 수 있다. 완만한 경사로가 두 개
층 높이의 공간을 돌아가며 하나로 엮었고, 한쪽 천창에서
온화한 빛이 떨어지는 열람실은 작은 구립 도서관의 중심
공간이 되었다.

　　　서울의 ‹압구정초등학교 다목적강당›과 ‹언북중학교
다목적강당› 프로젝트 역시 기존 학교 건축에 대한 불만에서
출발했다. 가장 평이하고 저렴한 경량철골로 만든 경사
지붕, 알록달록한 원색이 어린 학생의 창의성을 자극한다는
일반적인 가정 아래 산만하게 사용한 색들과 조각보처럼
덕지덕지 붙인 외벽 재료들. 대규모 공간에 어울릴 구조
형식에 관한 다양한 시도도 없이 학교 건물은 정말 이렇게만
만들어져야 할까? 학생들은 정말 이런 종류의 미적
감수성만을 좋아하는 걸까?

　　　두 다목적강당은 구조 형식에 관한 고민에서
출발해 반복과 수평성을 강조한 단순한 형태 어휘만으로

The ‹Maegok Library› is a relatively small library
owned by a district government, and did not have
the resources to enrich all its spaces. As with
most public architecture, it was a project which
required a process of selection and concentration.
Hence, we invested everything, all our efforts and
resources, into the reading space at the heart
of the library. What should be such an obvious
choice became the key line of reasoning behind the
inspiration of the ‹Maegok Library› design. The
smooth slopes would merge into one as they curved
around the two-story high reading room, enveloped
by tranquil top light in this central space of the
local library.

　　　The project of the ‹Apgujong Elementary
School Multipurpose Auditorium› and the ‹Eonbuk
Middle School Multipurpose Auditorium› in Seoul
was also inspired by our dissatisfaction with
existing school architecture, where the sloping
roofs are made of the most mundane and cheap
lightweight metal and walls are clumsily covered
with superfluous use of bright primary colors
based on the superficial assumption that this
would stimulate the creativity of young children.
Did school architecture really have to be like
this? Without any variation in approaching the
formal structure of such large-scale spaces and
considering what would suit them? Are students
truly only capable of appreciating this kind of
architecture?

완성했다. 비록 그 가운데 하나는 발주처의 요구에 따라
공모 당시의 구조 형식을 온전하게 유지할 수 없었지만,
구조적으로 단순한 형태가 지닌 아름다움을 구현하려는
원칙은 지켰다. 또한 재료를 선택할 때도 벽돌과 마루처럼
최대한 원재료의 색을 그대로 드러내는 마감재를 사용했고
체육관 내부는 차분한 계열의 무채색을 썼다. 채도가 강한
상징색은 프로젝트별로 홀에 딱 한 가지만 사용했다. 단순한
형태와 원재료의 물성 그리고 절제된 색상을 사용한 공간의
아름다움을 어린 학생들도 충분히 느끼고 이해하리라 믿었다.
기술과 비용이 받쳐주지 못하면 나아지기 어려운 형태나
디테일은 일단 미루어두고, 뜬금없는 색들을 여기저기에
무분별하게 사용해 머리를 아프게 하고 마음을 어지럽게
하는 일만 줄어도 우리의 학교 건축은 한결 나아질 것이다.

2018년 '한강건축상상전'의 전시작 ‹한강 가는 길›의
보행교는 자동차 위주의 도시 서울에 대한 전보림의
불만에서 동력을 얻은 프로젝트다. 한강과 지천이 만나는
합수부에 대한 도시적 제안이 주제였는데, 우리는 자동차
전용 도로가 도시와 강 사이를 막고 있는 한강 변의 구조적
결함을 극복할 수 있는 프로젝트를 제안했다. 기존의 연결
통로로 굴다리가 있지만, 아무리 기분 좋고 편안하게

The multipurpose auditorium began from our
problematizing structural form and was completed
solely with a simple formal language which
emphasized repetition and horizontality. While it
was impossible to maintain the original structural
form of our competition submission due to the
requirements of the commissioning body, we stood
by the principles of realizing an aesthetic
through a structurally simple form. The choice of
materials was also intended to maximally reflect
a finishing which would feature the natural tones
of the original materials, such as the bricks and
the wooden flooring, and a calm achromatic tone
was applied to the interior of the gymnasium. Only
one symbolic highly saturated color was selected
from each project. We believed that young students
would also sufficiently feel and understand the
beauty of a space which uses limited colors,
simple forms, and the materiality of natural
materials. We believe that we could easily improve
school architecture by just minimizing excessive
use of irrelevant colors, which causes headaches
and disturbance, as improvements on the form and
details are more difficult and require the support
of technology and money.

The foot bridge as an exhibition piece for the
2018 Han River Imagination Exhibition was a project
set into motion by the dissatisfaction of Borim Jun
regarding the automobile-centric city, Seoul. The
theme was an urban proposal about where the Han

압구정초등학교 다목적강당 (2018)
실내마감에서 바닥과 창의 기본 재료색은 그대로 두고
나머지는 한 톤으로 차분하게 통일하고자 했다.

APGUJONG ELEMENTARY SCHOOL MULTIPURPOSE
AUDITORIUM (2018)
In the interior, the primary material color of
the floor and window is revealed naturally,
and the others' colors are arranged with one
calm achromatic color.

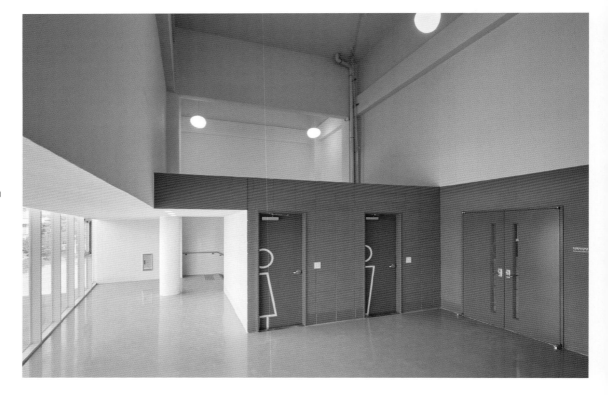

언북중학교 다목적강당 (2018)
천장의 높낮이를 달리해 공간 변화를 연출한 2층 홀은
주황색 계열의 상징색을 통해 공용 공간의 연속성을
유지하고자 했다.

EONBUK MIDDLE SCHOOL MULTIPURPOSE
AUDITORIUM (2018)
In the hall on the second floor, orange
symbolic colors are used. The architects
wanted to maintain the continuity of the
public space.

만들어도 굴다리는 굴다리일 뿐이다. 지하철이야말로
현재 서울 대중교통 시스템의 핵심이기에 한강과 가장
가깝게 맞닿는 옥수역 부근의 청계천 합수부를 대상지로
골랐다. 프로젝트가 너무 커서 할 일이 많다는 이승환의
또 다른 불만을 뒤로하고, 서로 1킬로미터 넘게 떨어진
옥수역과 서울숲을 잇는 보행길을 제안하자고 결정한 순간,
프로젝트는 이미 절반 이상 진행된 것이나 다름없었다.
불만의 힘이란 이렇게 강력하다.

river and its subsidiaries meet, and we proposed a
project which would overcome the structural faults
of the Hangang waterside where automobile only
roads obstruct the interstice between the city and
the river. Although connecting passageways exist,
these are underpasses, and after all, an underpass
is an underpass. We chose the site by Oksu Station
where the Chunggye stream joins the Han River,
as the metro has undoubtedly become the core of
contemporary Seoul public transport system. Setting
aside the dissatisfaction of Seunghwan Lee who
complained about how much work would be involved in
such a large project, the project was already half
done the moment we decided to propose a pedestrian
path connecting Oksu station and Seoul City Forest
which each lie over a kilometer away. This was the
strength of our dissatisfaction.

한강건축상상전 ‹한강 가는 길› (2018)
거대한 자동차도로로 가로막힌 한강변을 보행자에게
다시 돌려주자는 것이 제안의 핵심이다.

<THE PATHWAY TO HAN RIVER> FOR HAN RIVER
IMAGINATION EXHIBITION (2018)
The key message of the proposal is to return
the Han River, which is blocked by a huge
road, back to the pedestrians.

# 느림

# Slow

Essay Two

#건축사시험 #개소 #매곡도서관 #첫프로젝트 #설계공모
#공공건축 #디테일 #아름다움

↘ 사무실 벽에 붙이기 위해 페인트를
칠하고 있는 아이디알 건축사사무소의
로고.

아이디알IDR, Interdisciplinary&Integrated Design Research은 느리게
사는 전보림과 이승환이 마흔이라는 늦은 나이에
차린 건축사사무소로, 사무소 이름은 '아이디어idea'와
'아이디얼ideal'이라는 단어를 재료로 만들었다. 느림 자체가
삶의 목표는 아니지만, 느리게 살아감은 우리의 건축 여정을
설명하는 데 분명 핵심 언어다. 물론 느리다 혹은 빠르다
또한 어디까지나 상대적 기준이기는 하다. 만약 사무실을 연
시점이 서른다섯이었다면 그것은 이른 것인가, 늦은 것인가?
사실 우리가 하고 싶은 이야기는, 빠른지 느린지가 삶에서
그다지 중요하지 않다는 것이다.

　　대학원에서 만나 결혼하고 졸업한 뒤 아틀리에
사무실에서 실무 경험을 쌓아 건축사 자격증을 딸 때까지만
해도 우리 삶의 진도는 상당히 패스트 트랙이었다. 그러나
건축사 시험 직전 첫 아이가 태어나 세 돌이 될 때까지
아이를 직접 키우기로 하자 삶의 속도가 급격히 느려졌다.
첫 1년은 전보림이, 그다음 1년 반은 이승환이 아이를 맡는
식으로 번갈아 가며 육아를 전담했고 한 사람씩 돌아가며

116

#RegisteredArchitectExam #OpeningAFirm
#MaegokLibrary #FirstProject #DesignCompetition
#PublicArchitecture #Detail #Aesthetics

IDR, Interdisciplinary & Integrated Design Research
is an architectural firm opened by Borim Jun and
Suenghwan Lee at the ripe old age of forty, with
the words idea and ideal in mind. While slowness
is not necessarily a life goal, living slowly can
certainly be seen as a key word in explaining our
architectural journey. Of course, whether something
is late or early is always a relative standard.
If we had opened the architectural firm at thirty
five would that have been early or late? Actually,
the point we are trying to make is that whether
something is early or late is not that important to
living life.
　　After meeting in graduate school, getting
married and graduating, accumulating experience in
a small-scale architectural firm and acquiring our
registered architectural licenses, the progress we
were making in our lives seemed relatively fast.
However, the speed of such life became radically
slower the moment when our first child was born
right before the registered architect exam, as we
decided to bring him up by ourselves until his
third birthday. The first year Borim Jun, the
next year and a half Seunghwan Lee, took turns to
fully commit to childcare, while the other would
continue with architectural projects. As soon as

↗ The logo of the IDR
Architects, being painted
to be attached on the office
interior wall.

설계 작업을 이어갔다. 둘째가 태어나자 전보림이 다시
육아를 맡고 이승환은 대형 설계 사무소에서 일을 시작했다.
2년 뒤 사무실을 열 수도 있던 시점에 아이 둘을 데리고
오랜 바람이었던 유학길을 떠났다. 유학 기간에도 교대로
아이들을 돌보느라 공부 기간은 배로 늘어났고 직장 생활도
아이가 학교에 적응한 다음으로 미루다보니 또 늦어졌다.
그렇게 영국 런던에서 5년을 보내고 한국에 돌아왔을 때
어느새 나이는 마흔이었다. 그리고 전보림의 뱃속에는
셋째가 있는 상황이었다. 하지만 연이은 육아와 공부로
오랫동안 설계 일을 손에서 놓았다가 복귀하는 데 어려움을
겪었던 전보림은 셋째를 키울 때는 일을 놓지 않겠다고
다짐한 터였다.

셋째가 태어나기 꼭 닷새 전, 건축사사무소의
사업자등록을 했다. 셋째가 백일 무렵 첫 설계 공모에 지원해
당선되었고, 9개월 무렵부터 돌이 지나가는 넉 달이라는 시간
동안 우리는 갓난아기를 데리고 울산을 오가며 어렵게 구한
직원들과 함께 700평짜리 구립 도서관 설계 용역을 끝냈다.
말할 필요도 없이 전쟁 같은 시간이었다. 그렇게 태어난
아이디알의 첫 번째 프로젝트가 바로 ‹매곡도서관›이다.

부모가 첫아이를 키울 때 쩔쩔매듯이 ‹매곡도서관›

↖ 런던에서 귀국하기 직전, 처음으로
아이디알의 이름을 걸고 디자인해 런던
빅토리아앤드앨버트뮤지엄에 전시했던
‹컴플렉스, 레드 Complex, Red› (2014)
설치작품이다.

the second child was born, Borim Jun went back to
childcare and Seunghwan Lee started to work in a
large-scale design firm. Just when we could have
opened our own firm, two years later, we left to
study abroad with our two children, a long-desired
wish. Even during our time studying abroad, taking
turns to look after the children doubled the time
needed to study, and we further pushed getting a
job at a firm until after the children had adapted
to school, delaying things all the more. We spent
five years like this in London, and by the time
we had returned to Korea we had suddenly become
forty. At this point, Borim Jun became pregnant
with our third child. However, the difficulties
she had experienced in trying to return to work
from childcare or study made her resolved to stay
involved through this third pregnancy, and so the
company was founded.

Just five days before our third child was
born, we registered our architectural firm as a
business. When this third child was around 100
days old, we applied to and won our first design
competition. For the four months from when she was
9 months old to her first birthday, we traveled
back and forth between Ulsan and our office,
to complete the design project for the 2,100m²
district library with a carefully selected team of
staff. There is no need to explain how chaotic this
period was for us. Thus was born the first project
of IDR Architects, the ‹Maegok Library›.

↗ Just before returning
to Korea from London, IDR
Architects showed their
first installation work
called ‹Complex, Red› (2014)
displayed at the Victoria and
Albert Museum in London.

매곡도서관 (2017)
(위) 준공 직후, 보행자 진입 방향에서 바라본 모습.
(아래) 설계 공모 당시의 제출 이미지.

MAEGOK LIBRARY (2017)
(top) Right after completion, seen from
the pedestrian approach.
(bottom) Image submitted for the design
competition.

설계도 똑같았다. 공공 건축에 대해서는 아무런 경험이 없다
보니 공사비 내역서의 원가 계산 페이지가 무엇인지조차
몰랐다. 하나부터 열까지 모르는 게 너무 많았고, 그 가운데
절반은 알고는 있었지만 이해할 수 없는 것들이었다. 게다가
국가를 상대로 한 계약이니 뭐 하나라도 어기면 등본에 빨간
줄이라도 그어질 것 같아 겁이 났다. 어떻게든 정해진 기간에
일을 끝내야 했지만 동시에 첫 프로젝트였기에 대충할 수도
없었다. 모르고 부족한 상태에서 무언가를 주어진 시간에
만들어야 할 때 선택할 수 있는 방법은 오로지 쉬지 않는
것뿐이다. 거북이처럼 느렸지만 멈추지 않고 일을 계속했다.

전보림은 젖먹이 아기를 재워놓고 도면을 보았고
젖을 물리면서 전화를 했다. 이승환은 잠자는 시간만 빼고는
책상 앞에 앉아서 주야장천 일만 했다. 평평했던 의자가 미소
짓는 입처럼 푹 꺼질 정도로. 하지만 우리는 알고 있었다. 이
건물의 시공 과정에 감리로 참여할 가능성이 매우 낮으므로
그려달라는 대로 도면만 그려서는 좋은 건물이 나올 수
없다는 것을 말이다. 그런 상황에서 설계자가 할 수 있는
일은 도면을 최대한 꼼꼼하게 그리는 것과 사양을 자세하게
지정하는 것뿐이었다. 그래서 아둔할 정도로 도면을 그리고
그리고 또 그렸다. 업체에 견적서를 보내달라는 전화를 하고

↳ 전보림의 용기와 담당 주무관들의
배려로 프로젝트를 진행하는 내내
막내인 셋째 아이를 데리고 울산에서
열리는 회의에 참석하며 일을 했다.

In the same way that novice parents have a hard
time, the design process for the ‹Maegok Library›
was quite challenging. Without any experience
regarding public architecture, we didn't even
understand how the production cost was estimated.
There were so many things we didn't know and even
if we did, we couldn't fully understand them.
Moreover, having signed a contract with the state,
we were scared that the slightest deviance would
end up with red marks all over personal records.
While it was imperative that we finish the project
in the designated time, we couldn't just wing it
as it was our first project. The only choice for
someone who has to make something in a designated
amount of time, lacking in knowledge and competence
is simply to forego rest. As slow as turtles as we
were, we continued inexorably on, without stopping.
Borim Jun would look at the drawings as soon
as the child fell asleep, and would take calls
while breastfeeding. Seunghwan Lee would work
away day and night, except for the time needed for
sleep. To the point that the flat surface of the
chair deflated in the form of a smiling mouth. We
knew that it was improbable that we would be able
to participate in supervising the construction
process, and hence, that good architecture could
not result from the standard set of drawings
and documents started in the contract. In this
situation, the only thing that the architect can
do is to include as much detail as possible to the

↗ With sincere consideration
of the co-workers, Borim Jun
attended meetings in Ulsan
with her youngest child,
throughout the project.

또 했다. 납품하면서 세어보니 넉 달 동안 그린 건축 도면과
구조 도면은 220장이었고 견적서는 70장이었다.

영영 끝날 것 같지 않던 ‹매곡도서관› 프로젝트는 끝이
났다. 하지만 셋째는 겨우 돌이 갓 지난 아기였다. 게다가
급하게 꾸리기는 했어도 프로젝트를 하나 끝낸 사무실은
제법 그 꼴을 갖춘데다가 먹여 살릴 사무소 식구까지 하나
늘어난 상태였다. 이제는 프로젝트 수주로 먹고 살아야 했다.
그래서 프로젝트가 없는 작은 설계 사무실이 그러하듯 우리
역시 당연한 수순처럼 설계 공모전에 도전하기 시작했다.
‹매곡도서관›이 당선되었으니 또 도전하면 될 것도 같았다.
그러나 그런 일은 그리 쉽게 일어나지 않았다.

약 10개월 동안 여섯 개의 설계 공모에 도전했다.
하지만 결과는 모두 낙방이었다. 아이디알의 설계 실력이
문제일까, 심사위원의 안목이 문제일까? 이도 저도 아니면
그저 다른 힘의 문제일까? 납득할 수 없는 결과들에 지쳐갈
무렵, 교육청에서 발주하는 체육관 격의 다목적강당
설계 공모에 두 개가 한꺼번에 당선되었다. 원래는 전자
입찰로 처리하던 소규모 설계 용역이었는데 사업 초기라
응모 업체가 많지 않았던 것이다. 연이은 낙방에 기운이
빠져서 그야말로 최소한의 내용만 갖춰 제출했는데

ꜛ ‹언북중학교 다목적강당›의 설계
공모 제출 이미지.

ꜛꜛ ‹압구정초등학교 다목적강당›의
설계 공모 제출 이미지.

drawing set, and specify each detail to perfection.
This was how we ended up producing drawings over
and over again, to the point it felt a little
offensively explicit. We called the contractors
over and over again demanding estimate after
estimate. During the final delivery, we counted
220 architectural and structural drawings and 70
estimates over 4 months.

The ‹Maegok Library›, a seemingly
interminable project eventually came to an end.
Our third child had just had her first birthday.
The office, haphazardly put together, yet having
completed an entire project, had started to take
shape, with an additional mouth to feed. Now,
the firm had to survive on commissioned projects.
Hence, like all small architectural firms without
projects, we also followed the conventional wisdom
and launched ourselves into the challenge of design
competitions. Since we had been selected for
‹Maegok Library› who was to say we wouldn't rise
to the challenge a second time? This didn't happen
right away.

For around 10 months, we applied to six
design competitions. We failed at all of them.
Was the problem the design competence of IDR
Architects, or the preference of the judges? If
it wasn't either, then was there a different
issue, like some sort of power? When we were just
starting to become exhausted by failure, we were
nominated for two design competitions at once for

ꜛ Image of <Apgujong
Elementary School
Multipurpose Auditorium>
submitted for the competition.

ꜛꜛ Image of <Eonbuk
Middle School Multipurpose
Auditorium> submitted for
the competition.

오히려 당선되었다. 이렇게 아이디알의 프로젝트 여정은
‹매곡도서관›이라는 공공 건축으로 시작해 이번에는 학교
건축이라는 다른 범주의 공공 건축으로 이어지게 되었다.

　큰 범위에서는 같은 공공 건축이지만, 발주처가
교육청인 학교 건축은 완전히 다른 신세계였다. 건축 설계
분야만 제대로 하려고 해도 알아야 할 지식이 끝도 없이
쏟아지는데, 거기에 분야의 특수성까지 더해지니 도무지
정신을 차릴 수가 없었다.

　그 와중에 첫째와 둘째는 한국에 와서 2년 동안 다니던
대안학교를 그만두고 홈스쿨링을 시작하게 되었다. 영국
생활에서 익숙해진 영어가 공부 언어로 그리고 또래와의
소통 수단으로 확고히 자리 잡아 한국어로 공부하는 것이
불편하고 재미없었던 듯하다. 결국 집에서 부모인 우리가
아이들을 직접 가르치기로 했다. 책 읽어달라, 놀아달라
보채는 셋째 아이를 한쪽 무릎에 앉히고 두 아이를 가르치는
말도 안 되게 산만한 홈스쿨링이었다. 그렇게 1년을 보내고
우리는 이 방식으로 계속하는 것은 무리라고 깨닫고
아이들을 미국 웹사이트 기반의 홈스쿨링 프로그램에 등록해
공부하도록 했다.

　그동안에도 교육청 프로젝트는 진행되었다. 서로 다른

multipurpose auditoriums which were gymnasiums,
commissioned by the Office of Education. Considered
as smaller-scale projects, these buildings had been
commissioned through digital bidding, but when
the process was switched to design competition,
there were not many firms that participated.
We had been selected despite having submitted
just the bare bones, due to our exhaustion with
consecutive failures. As such, IDR's trajectory
of projects started with the public architecture
project of ‹Maegok Library› and continued with the
public architecture of another kind, namely school
architecture.

　At a glance it was the same public
architecture, yet school architecture with the
Office of Education as a client was an entirely new
world. We felt overwhelmed by all we had to learn
about the genre, in addition to all we had to learn
about design itself.

　In the midst of this, the two older children,
who had been going to an alternative school for
the two years since they had returned to Korea,
left their schools to start homeschooling. It
seemed that they found it uncomfortable and
uninteresting to study in Korean, as English had
been firmly established in their minds as the
language of studying and communicating with their
peers from their time in the UK. Eventually, we
as the parents, decided to teach them ourselves
at home. With the youngest child on one knee, who

언북중학교 다목적강당 (2018)
반복적인 모듈을 통해 형태와 빛의 리듬이 만들어내는
아름다움을 구현하고자 했다.

EONBUK MIDDLE SCHOOL MULTIPURPOSE
AUDITORIUM (2018)
Through repetitive modules, the architects
tried to realize the beauty created by the
rhythm of form and light.

언북중학교 다목적강당 (2018)
창은 '빛의 띠'가 되었다.
부드러운 자연광이 실내에 깊숙하게 들어오도록 했다.

EONBUK MIDDLE SCHOOL MULTIPURPOSE
AUDITORIUM (2018)
The window looks like a band of light. It
allows natural light deep inside the room.

두 학교의 다목적강당을 설계하는 것이었지만 교육청의 시설 표준 요구 사항은 거의 동일했으므로 중복되는 설계 내용이 많았다. 그러나 전에는 해본 적이 없는 것들이어서 역시 쉽지 않았다. 교육청이나 업체에서 참고하라고 보내준 도면이 있었지만, 그 디테일을 그대로 적용하고 싶지 않았다.

군이 일하면서 말로 표현하지는 않지만, 아이디알이 건축 설계를 통해 표현하고자 노력하는 것 가운데 하나는 아름다운 디테일이다. 기본 도면과 표준 디테일로만 지어지는 건축물을 우리는 전혀 좋아하지도 않을뿐더러 작품이라는 말을 붙이고 싶지도 않다. 그런데 너무나도 많은 건물이 심지어는 어느 정도 이름이 알려진 건축가나 유명 사무실에서 설계한 건물조차도 그렇게 지어지는 경우가 많다. 특히 공공 건축이 그렇다. 여러 이유가 있겠지만 아마도 표준 디테일이 아닌 자신만의 디테일을 만드는 일에 시간과 노력이 많이 들기 때문일 것이다. 업체에 기술 자문을 구하려 해도 공공 기관 납품을 주로 하는 영세 업체는 표준 디테일이 아닌 다른 디테일은 시도할 생각조차 해본 적이 없는 경우가 허다하다. 〈매곡도서관〉의 루버 시공을 위한 디테일 도면을 그리는 과정도 일일이 참고서를 들춰가면서 그야말로 맨땅에 헤딩하는 심정으로 시작해야 했다.

continuously asked us to read her books and play with her, we started the unimaginably chaotic homeschooling of two children. After a year of this, we realized that we had expected too much of ourselves, and registered them to study with an online American homeschooling program.

The Office of Education project continued throughout this. While it was required that we design the multipurpose auditoriums of two different schools, a lot of the design content overlapped with almost identical standard facility requirements outlined by the Office of Education. Without any prior experience, it wasn't easy. The Office of Education and the related firms sent over drawings we could refer to, yet we didn't want to simply copy and paste these details.

While we never really put this into words while we work, one of the things that IDR strives to express through architectural design is aesthetically pleasing detail. Not only do we find architectural works built solely with basic drawings and standard details entirely unappealing, we would not even define them as architectural works. Nevertheless, so many buildings, even buildings designed by relatively well-known architects or famous design firms are often built in this way. Public architecture is even more so. Of course, there must be their reasons, yet it is probably because of the time-consuming and labor-intensive nature of making their own unique details

↗ Detail view of the louvers of <Maegok Libary>.

아무리 건축가가 여러 분야의 지식을 섭렵해 종합하는
직업이라고는 해도 자재의 물성 등을 고려한 해당 분야
전문가의 기술 정보 지원은 필요하다. 그런데 그런 지원도
없이 발주처도 자재업체도 시공사도 전혀 달가워하지 않는
새로운 디테일을 그리는 것은 결코 만만한 일이 아니다. 건물
디테일이라는 것이 뭐가 그리 대단하고 꼭 별난 디테일이
있어야만 좋은 건물이 되는지, 그만한 노력을 들일 가치가
있는 것인지 싶을지도 모르겠다. 그러나 서로 다른 재료와
재료가 만나는 부분이 아름답지 않은데 어떻게 건물이
아름다울 수 있겠는가. 남들보다 한 발짝 더 나가는 일은
분명 고되지만, 그 한걸음이 있어야 '남다른 아름다움'이
만들어진다고 우리는 믿는다.

매번 그렇게 일을 하다 보니 늘 기진맥진이다. 한
프로젝트가 끝나면 잠시 쉬어야만 겨우 다른 프로젝트에
집중할 기운이 생긴다. 일이 끊어지지 않게 하려면
프로젝트가 끝나기 전에 다른 프로젝트 공모전을 시작해야
한다는 것이 업계의 정설이다. 하지만 우리에게는 어림없는
일이다. 나이는 벌써 마흔다섯인데 사무실을 열고 5년여 동안
준공작은 단 네 곳뿐이다. 이미 이렇게 늦었는데도 아이들을
직접 키우느라 언제나 둘이 힘을 합쳐도 한 사람 몫밖에

that differ from standard details. Even if we tried
to request technical advice from contractors, the
small private firms which normally operate on
delivering for public organizations have often
never even thought about attempting details that
differ from a standard design. The process of
producing the construction drawings for the louvers
in the <Maegok Library> project made us feel like
we were starting from scratch, utilizing every
reference we could find.
    However much an architect may be a career
which must search far and wide, and comprehensively
put together knowledge from a variety of fields,
we do require technical information support from
specialists when considering the properties of
certain materials. Hence, it was no easy matter
to draw up a set of new details that the client,
the suppliers, and the contractor feel less
enthusiastic about. One may question the value
of architectural detail, whether it is important
to have unique detail to make it a good piece of
architecture, whether it is worth putting so much
effort in. Yet, how could one call a building
beautiful if the intersection between one material
and another is not aesthetically pleasing? While it
was certainly a difficult task to go a step further
than others, we believed that it was this step that
would create "a unique beauty."
    It's a exhausing process. When one project
ends, we always take a break before regaining

언북중학교 다목적강당 (2018)
반투명의 폴리카보네이트 패널과 메탈 루버를 덧댄
창으로 만들어진 '빛의 띠'.

EONBUK MIDDLE SCHOOL MULTIPURPOSE
AUDITORIUM (2018)
The 'Light Stripes', composed of translucent
polycarbonate panels and windows with metal
louvers on them.

못하니 더 느리다. 동기들의 사업 경력은 우리보다 10여 년은 앞서 있다.

그러나 느리게 가는 삶이 그다지 나쁘지는 않다. 삶의 풍경을 훨씬 더 섬세하게 느끼고 음미하며 살게 되기 때문이다. 누군가 우리에게 일을 위해 사는지, 아니면 살기 위해 일하는지 묻는다면 우리는 그 어느 쪽도 아니라고 대답할 것이다. 우리에게 삶은 일하며 가족과 함께 살아가는 지금 이 순간의 연속이다. 다른 사람들이 어떤 속도로 무엇을 성취하며 사는지 우리 삶에서는 전혀 중요하지 않다. 그저 우리에게 소중한 가치에 집중하며 살아가려 한다. 그래서 아이들이 자라는 매 순간에 충실하고 우리가 설계하는 건물 하나하나에 정성을 다하며 이따금 글을 쓰며 느릿느릿 살아가고 있다.

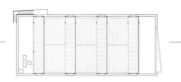

ㄱ 〈언묵중학교 다목적강당〉 지붕평면도.

the strength to start another. Yet, it is widely accepted in the industry that one must start out on another competition before finishing off a project in order to have a continuous flow of work. For us, this was impossible. We are already 45, and we only have four completed projects since opening the firm five years ago. Not only are we quite late, bringing up the children by ourselves means that we are that much slower as we can only do the work for one even when we join forces. The business acumen of our former colleagues is already a decade ahead of ours.

Nevertheless, there is not much at fault with this slow pace of life. It forces us to feel and to appreciate with greater detail, the landscapes of our lives. If someone were to ask us if we work to live, or if we live to work, we would say that it is neither. For us, life is working, living with family, and a continuation of present moments. It is not at all important to us to learn about what others accomplish in their lives. All we wish to do is live and concentrate on the values we hold dear. Hence, this is how we live a slow life, cherishing the moments of our children growing up, dedicating ourselves to each and every one of the projects we design, and occasionally writing about something we feel like.

↗ Roof plan of the <Eonbok Middle School Multipurpose Auditorium>.

# 공공

# Public

Essay Three

2014년 5년 동안의 런던 생활을 끝내고 돌아왔을 때
우리에게는 마땅히 계획이랄 게 없었다. 몇 달 뒤
사무소를 열기는 했지만, 딱히 할 일도 없었다. 빈둥거릴
바에는 설계 공모라도 지원해보자며 시작한 프로젝트가
〈매곡도서관〉이었고, 우여곡절 끝에 첫 도전에 당선이라는
행운을 안았다. 지금 생각해도 놀라운 일이었다. 그렇게
시작한 공공 건축은, 무엇보다 우리 둘 다 민간 영역에서
일을 만들어내는 재주가 없어 그렇기도 하지만, 5년이 지난
지금 우리의 건축 작업을 관통하는 주제가 되었다.

　　시작은 분명 우연에서 비롯되었다. 그러나 해놓고
나면 필연처럼 느껴질 때가 있듯이 마치 지금까지의 경험이
공공 건축을 마주하는 지금의 우리 태도를 만들어낸 것 같다.
2007년 둘이 같이 다니던 설계 사무소를 휴직하고 처음
서로 머리를 맞대고 참여해 은상을 받았던 〈공중화장실
현상공모 계획안〉이나 공공 역할이 두드러진 런던에서의
유학 생활이 그렇다.

　　공공公共의 사전적 의미는 모두에게 열린公 자원을

By the time we returned from our five year sojourn
in London in 2014, we didn't really have what you
might call a plan. We did open the firm a couple
of months later, but there wasn't any work to be
done. The project we had applied ourselves to, with
the thought that it would be better to work on a
design competition than lounge about the office was
the <Maegok Library>, and with some ups and downs,
we were lucky enough to win a competition with our
first attempt. It's quite something to think of,
even now. Public architecture, which we started in
this way, has become a running consistent theme in
our most recent architectural projects, five years
on, perhaps because both of us aren't great at
making projects happen in the private sector.
　　We started this project completely by chance.
Yet, just like one is prone to feel that after
completing something it was destiny, it feels as
if our experience to this day formed our current
approach in dealing with public architecture.
This is true of the time that we won the silver
award for the proposal of 'Public Toilet Design
Competition', the first case in which we put our
heads together after taking extended leave from the
architectural design firm that we both worked at in
2007, or the life we led in London which was highly

함께 共 나누는 일이다. 그러므로 우리는 건축 설계의
가치를 가장 효과적으로 전달할 수 있는 길이 공공 건축에
있다고 믿는다. 그리고 현실에서의 중요성만큼 공공 건축을
제대로 설계하고 짓는 것이 건축가가 맡은 중요한 사회적
역할이라고 생각한다.

우리가 공공 건축을 설계할 때 가장 중요하게
생각하는 점은 그 공간의 근본을 다시 생각해 다수가 함께
향유할 수 있는 즐거움을 담는 것이다. 공공 건축의 본질적
역할을 고민하고 이를 건축으로 구현하여 모든 사람이
가치를 두루 누릴 수 있게 만들어간다. 그리고 공공 건축이
지역 사회와 유기적 방식으로 조화를 이루며 그 안에 사는
사람들과 함께 나이 들어갈 수 있기를 기대한다. 그 과정이
험난하더라도 우리가 공공 건축에 끊임없이 관심을 기울이는
이유는 잘 만든 공공 건축이 더 많은 사람에게 윤택한 삶을
누릴 수 있게 하고 건축 설계에 대한 인식을 바꾸는 계기를
제공하기 때문이다.

이런 프로젝트를 할 때 우리가 의존하는 것은
역설적으로 매우 사적인 기억과 경험이다. 가장 높은
수준의 공공성은 개인의 희생을 강요하지 않는다고 우리는
믿는다. 결국 모든 공공 공간은 최종적으로 사적인 영역에서

public in nature.
The dictionary definition of public is the
act of sharing resources which are open to all.
Hence, we believe that the path to most effectively
communicating the values of architectural design
lie in public architecture. We also believe that
an important social role of the architect is to
properly design and build public architecture which
reflects its significance in reality.
What we think of as most important when
designing public architecture is to rethink the
fundamental nature of a space, so that the space
can embrace a sense of joy which can be appreciated
by many. We carefully consider the fundamental
role of public architecture and translate this
into architecture, so that all people can equally
benefit from its values. We also hope that public
architecture can exist harmoniously with the local
community in organic ways, and that it can age
with the people who live there. We continue to
pay attention to public architecture despite this
bumpy process, as well as build public architecture
that can allow even more people to benefit from an
enhanced quality of life, and can trigger a change
in awareness about architectural design.
What we depend upon when carrying out such
projects is, ironically, extremely private memories
and experience. We believe that publicness executed
at the highest degree does not necessitate the
self-sacrifice of the individual. Ultimately, all

소비된다. 5년이라는 영국 생활에서 우리는 공공이라는
완충재를 통해 개인이 보호받고 어울려 사는 도시를
경험했다. 특히 도시와 건축이 어우러져 만들어낸 도시
공간이라는 공공재는 공공성이 낳은 가장 값진 결과임을
깨달았다. 그러한 사적인 기억과 경험에는 주관적 진실에
기반한 통찰력이 깃들어 있다고 생각한다.

많은 공공 건축은 불특정 다수가 이용하므로
사회 통념에 기댄 공허한 가정을 바탕으로 설계를
시작하는 경우가 많다. 전지자全知者를 꿈꾸는 설계자의
오류에서 벗어나기 위해 우리는 각자 경험의 조각들을
서로 주고받으며 건축적인 살을 조금씩 붙여나갔다.
‹매곡도서관›의 통합된 열람실은 불편의 경험이 새로운
대안을 열어준 경우다. 또한 비록 실현되지는 못했지만,
울산 북구의 ‹호계문화체육센터 계획안›에서 수영장을
내려다보도록 설계한 휴게 공간은 런던에서 살 때 즐겨 찾던
동네 체육센터 로비의 유쾌함을 떠올리며 진행했다.

공공 이익과 사적인 만족은 동시에 충족되기
어려운 역설적 관계에 있다. 따라서 이런 공공 공간이 사적
경험을 충실히 담아낼 수 있도록 우리는 개인이 어떻게
주변 공간을 전유화專有化하는지 주목한다. 말하자면 모든

↘ 런던의 가장 공적인 공간인 잔디밭
광장. 사람들은 자연스럽게 이 공간을
자신만의 방식으로 즐긴다.

public spaces are consumed in the private domain.
During our five years in the UK, we experienced a
city in which individuals, protected by the buffer
zone of the public, live together. In particular,
We realized that the urban space established with a
sense of harmony between the city and architecture,
while being a common resource, can be the most
valuable result of publicness. We believe that
these personal memories and experiences contain
insights that are based on subjective truths.

As many public buildings are used by
undefined mass of people, often the design process
is set on a vacuous assumptions based on generic
social norms. To walk way from the designer's
mistake of being the god who knows everything, we
built up architectural ideas by sharing our own
bits of experiences.

The integrated reading room of the <Maegok
Library> is a case where the experience of
inconvenience opened up new alternatives. Also,
while it remains unrealized, we looked back on the
pleasure we derived from the local sports center
in London while developing the competition entry
for the <Hogye Cultural Sports Center> in Buk-
gu of Ulsan by designing a lounge overlooking the
swimming pool.

Public interest and personal satisfaction form
a paradoxical relationship in that it is difficult
to simultaneously realize both. To understand how
such public spaces can fully embrace the personal

↗ A grassy field, one of
London's most public spaces.
People naturally enjoy this
space in their own way.

사람을 위한 공간이지만, 동시에 나만을 위해 존재하는
것 같은 공간이 목표다. 그런 상황을 일으키는 물리적
조건을 면밀하게 관찰하고, 이를 건축적으로 재구성하기
위해 노력한다. 우리가 발견한 것은 균질한 공간의 밋밋함
대신 비균질적 조건을 가진 공간이 제공하는 입체적이며
다중적인 가능성이다. ‹매곡도서관› 열람실은 아홉 곳의 바닥
높이가 서로 다르게 구성되어 있으며 길이 방향을 따라가며
끊임없이 변하는 단면을 가지고 있다. 중정 옆의 계단식
좌석과 그 양쪽 창을 바라보는 자리, 천창 바로 아래의 밝은
테이블, 두 열람실을 바로 연결하는 가운데 계단 뒤쪽의
어둑어둑한 틈새 공간 등 서로 다른 공간적 덩어리감, 외부
전망, 그리고 빛의 표정을 가진 자리들은 이런 구조를 통해
만들어진다. 그리고 그 자리마다 자기만의 세계에 빠진 듯
편안하게 책을 읽는 사람들을 만날 수 있다.

또한 우리는 공공성의 위계에 따라 공간을 구분할
수 있는 그 가능성에 주목한다. 공공 건축은 대부분 누구나
자유롭게 출입할 수 있는 공간과 특정한 자격이나 절차가
요구되는 공간으로 구분된다. 또한 출입 제한이 없는
공간이라도 건물의 핵심 기능을 담당하는 곳과 부차적인
역할을 담당하는 공간으로 나뉜다. 이때 자유롭게 접근할 수

133

experience, we focus on how the individual might
appropriate their surrounding environment. The goal
is to create a space which feels like it exists
just for me, while being a space for everyone
at the same time. We try to closely observe the
physical conditions in which such situations occur,
and to recreate this through architecture. What
we discovered is how spaces with heterogenous
condition provide multiple, rich possibilities,
opposed to the blandness that homogenous spaces
offers. The reading room of the <Maegok Library>
is composed of nine different floor levels with
sections which continuously transform as you follow
in its path. It is the structure which creates such
heterogenous spatial volumes, the outside views,
the place characterized by lights such as the
stepped seating next to the courtyard, the spaces
which look onto the windows at each side, the table
underneath the top light, the shadowy niche spaces
at the back of the stairs which directly connect
the two reading rooms, etc. And at each place it is
possible to encounter people comfortably reading,
deeply immersed in their own worlds.
    We also focus on the possibility of
differentiating space according to the hierarchy of
publicness. Most works of public architecture are
differentiated through spaces with open access to
anyone, and spaces which require a certain status
or process. Also, spaces with open access have
spaces which manage the central function of the

134

매곡도서관 (2017)
열람실 안 창가에는 다양한 형태의 좌석이 구성되어 있다.

MAEGOK LIBRARY (2017)
There are various types of seats near the
window in the reading room.

135

매곡도서관 (2017)
건축가는 열람실이 모두를 위한 공간인 동시에
자신만의 공간으로 느껴지길 바랐다.

MAEGOK LIBRARY (2017)
The architects wanted the guests to feel that
the reading room is both a space for everyone
and his/her own space.

매곡도서관 (2017)
1층 어린이 열람실 전경. 2층 일반 열람실과
하나의 공간으로 연결되어 있다.

MAEGOK LIBRARY (2017)
The view of the children's reading room on
the first floor. It is connected to the
general reading room on the second floor.

있는 공간 그리고 핵심 기능을 담당하는 공간을 풍요롭게
만들수록 건물 전체의 공공 가치는 커진다. 〈매곡도서관〉의
다소 초라한 로비, 이와 대비되는 풍요로운 열람실이 대표적
예다. 경기도 김포시 〈풍무도서관 계획안〉의 열람실 또한 큰
틀에서는 비슷한 의도를 가지고 설계했다. 열람실 한가운데
커피 향이 번지는 활기찬 카페 거리가 있는 '건물 속 건물'
같은 도서관이다. 두 곳 모두 도서관이 공부보다는 책을 읽고
지식을 확장하는 공간이어야 한다는 평소 신념을 바탕으로
자유롭고 느슨하게 만들었다.

〈호계문화체육센터 계획안〉은 수영장을 바라보는 휴게
카페와 건축적 연출에 많은 공을 들인 프로젝트다. 헬스장과
수영장 이용자, 그들을 기다리는 사람에게 빛이 충만한
수영장과 그 너머의 풍경을 바라보며 여유로움을 느끼게
하고 싶다는 의도에서 비롯되었다.

아직은 실험 단계에 머물러 있지만, 우리는 특정한
용도 없이 다소 모호하면서도 특징적이며 때로는 시적
감각을 불러일으키는 공간에도 관심이 있다. 이런 열린
가능성을 지닌 공간이 공공성을 띠게 되었을 때 사용자가
만드는 긍정적이고 우발적인 사건은 모든 것을 예측할 수
없는 건축가의 한계를 보완해줄 뿐만 아니라 유연한 공간

↖ 〈풍무도서관 계획안〉 1층 열람실
투시도. 공원의 길이 실내로 이어진
듯한 느낌이 들도록 계획하였다.

↖↖ 〈풍무도서관 계획안〉 2층 북카페
투시도. 도서관의 중심 공간이 마치
카페와 같이 편안한 독서 공간이 되길
바랐다.

building, and spaces which take on the supporting roles. Here, adding a sense of richness to the design of open access spaces and the central function spaces increases the overall public value of the entire building. A good example is the relatively humble lobby and the contrasting rich spaces of the reading room in the <Maegok Library>. The reading room from the competition entry for <Pungmu Library> in Kimpo, Geyonggi-do also had the same broad intention. It was planned to be a library with a "building within a building" in which a lively café street, with fragrant scents of coffee, could be found in the reading room. Both of these spaces were designed to be free and loose, based on our everyday conviction that a library should be a place to read books and extend one's sense of knowledge rather only for study.

The competition entry for <Hogye Cultural Sports Center> was a project where we spent a lot of care and attention for the leisure café overlooking the pool and the architectural sequence around it. This came from the motive that we wished to give a sense of leisure to people using the fitness center and swimming pool and those waiting for them as they looked across the swimming pool full of light and the landscape beyond.

While it still remains in its experimental phase, we are also interested in spaces with no particular purpose, which seem slightly ambiguous yet unique, and even at times feel quite poetic.

↗ The competition entry for <Pungmu Library>, the view of book café on the second floor. The architects hoped that the center of the library would be a comfortable reading space like a cafe.

↗↗ The competition entry for <Pungmu Library>, the view of the reading room on the first floor. The first floor is designed to feel like connected to the park outside.

이용을 통해 공공 건축의 공공성을 더욱 확대해줄 것이다.
〈매곡도서관〉 어린이 열람실의 특징인 높이가 서로 다른
세 개의 단과 〈강동문화센터 계획안〉의 입체적 아트리움은
수직 수평의 동선 문제를 해결함과 동시에 이런 가능성을
실험하기 위한 접근이었다. 비록 미미한 시도에 불과하지만,
머지않은 미래에 조금 더 합당한 용도와 스케일을 가진
프로젝트에서 그럴 듯하게 구체화되기를 기대해본다.

ㄱ 〈호계문화체육센터 계획안〉 수영장을
볼 수 있는 2층 카페 투시도. 체육시설인
동시에 지역 사회의 구심점이 될 수
있도록 모두가 쉽게 접근할 수 있는
공공공간을 만들고자 했다.

Spaces with such open potential, when defined
for public use, result in users creating positive
spontaneous events, which not only complements
the limits of the architect, who cannot predict
everything, but can also extend the public nature
of public architecture through the flexible use of
space. The children's reading room characterized
by the three different levels at the <Maegok
Library>, as well as the spatially dynamic atrium
of the competition entry for <Gangdong Cultural
Center> were both proposed to resolve issues of
vertical and horizontal circulation while becoming
an approach to experiment with such potential.
While these were modest attempts, we hope that
they will be realized in the near future with more
appropriate conditions and situations.

ㄱ The competition entry
for <Hogye Cultural Sports
Center>, the view of café on
the second floor where one
can overlook the swimming
pool. The goal was to create
a public space that is easily
accessible to everyone as
well as a sports facility.

# 배경

# Background

Essay Four

#계양3동주민센터 #명동센트럴 #풍무도서관
#호계문화체육센터 #매곡도서관 #다목적강당
#매개자 #균형

ㄴ 피츠행어저택 서재의 창문.
이 사진에는 세 가지 요소가 공존한다.

건축가로서 자신의 작업을 관통하는 하나의 주제를 갖는다는
것은 동전의 앞뒤처럼 이중적인 면이 있다. 예술가로서
건축가의 세계를 일관성 있게 보여주기도 하지만, 작업의
다양성을 스스로 가두는 족쇄가 될 위험성도 동시에 있다는
말이다. 그렇기 때문에 객관적인 비평가의 시선을 통하지
않고 건축가 스스로 자신의 건축관을 드러내는 건 쉬운
일이 아니다. 프로젝트마다 겉으로 드러나는 건축 어휘의
유사성에 현혹되지 않고, 그 기저에 깔린 대상과 마주하는
태도에서 일관된 이야기의 실마리를 찾는 지혜가 필요하다.
　　우리는 한 장의 사진에서 이 조심스러운 이야기를
시작한다. 사진은 전보림이 런던에서 공부할 때 스튜디오
답사로 방문한 존 손 John Soane의 피츠행어저택 Pitzhanger
Manor House에서 찍은 것이다. 존 손은 런던 영국은행을
설계한 신고전주의 건축가로 영국 건축계에서는 흥미로운
컬렉션으로 가득한 존손뮤지엄 Sir John Soane Museum 덕에 꽤
유명한 인물이다. 그가 짓고 살았던 집 서재에서 정원이

#Gaeyang3dongLocalOffice #MyeongdongCentral
#PungmuLibrary #HogyeCulturalSportsCenter
#MaegokLibrary #Mediator #Balance

As an architect, it has both pros and cons to have
a single theme throughout the projects. While it
can be used to showcase the consistency in the
architect's works like an artist, it also carries
the risk of a self-imposed prison limiting the
diversity of work. Consequently, it is not easy
for an architect to take the lead in showcasing
one's work without depending on a view of a critic.
This requires the wisdom of remaining unseduced by
the similarities in the architectural expressions
in each project, and to seek out small outlets
from the architect's underlying attitude towards
projects.
　　We cautiously begin our story with a single
photograph. The photograph is from when Borim
Jun was studying in London, and it is a picture
of the Pitzhanger Manor House by John Soane,
taken during a studio site visit. John Soane, a
neo-classical architect who designed the Bank
of England in London, is quite well known in
the British architectural field due to the Sir
John Soane Museum filled with an interesting
collection. This photo, which features a window
facing the garden is taken from the study, in the
house he built and lived in, and at first seems
to have nothing in particular. Yet, upon closer

ㄱ A window detail at the
reading room of Pizhanger
Manor House. one can see
that three elements co-exist
within the same scene.

보이는 창을 찍은 이 사진은 처음 봤을 때는 딱히 특별한
것이 없어 보인다. 그런데 이 사진을 자세히 보면 장면 안에
세 요소가 공존하는 것을 알 수 있다. 하나는 이 집의 외부
공간, 정확히는 발코니와 그 너머의 정원이고 다른 하나는
이를 바라보고 있는 시선의 주체다. 그리고 마지막으로 이 둘
사이를 이어주는 창이 있다. 창의 많은 부분이 투명 유리로
되어 있어 처음에는 그 존재가 바로 느껴지지 않는다. 창은
원래 그 너머를 보기 위해 디자인된 것이기 때문이다. 그러나
사진의 이런 첫인상의 순간이 지나가면 다른 것이 눈에
들어오기 시작한다. 200년도 더 된 집이기에 당시의 기술과
양식에 따라 유리가 잘게 분할되어 있고, 곳곳에 과하지 않은
몰딩이 제 위치에 자리 잡고 있다. 가로 부재와 세로 부재의
두께가 대조를 이루는 창살, 측면에 점잖게 마감된 덧창 패널,
적절한 크기와 형태로 다소곳이 붙어 있는 문 손잡이 같은
것도 차츰 시선을 끈다. 말하자면 이 사진에서 창은 주인공의
위치에서 한발 뒤로 물러나 시선의 주체와 대상을 연결하는
역할을 한다. 그러면서 섬세한 디테일로 그 존재감의
균형점을 잃지 않는다.

아이디알의 건축은 이와 닮았다. 전면에 화려하게
나서기보다는 마치 배경처럼 서로 다른 것을 이어 주고

observation, one can see that three elements co-
exist within the same scene. One is the external
space of the house, more accurately the balcony and
the garden beyond, while the other is the person,
gazing at the garden, looking out. Finally, there
is the window that connects these two together. As
a large part of the window is made of glass, at
first, it easily goes unnoticed. This is because
a window is fundamentally designed as a device to
see beyond. Yet, once this first impression of
the photo slowly fades away, other elements start
to catch one's eye. In this house, which is more
than two centuries old, the window panels were
designed according to the technology and style of
its time, with adequate moldings at appropriate
locations. The thickness of the vertical and
horizontal architectural elements contrast through
the cross ribs of the window, the wooden panels
gently attached to the side, the window handle
calmly installed with proper size and form, slowly
draw one's attention. In other words, the window in
this photo plays the role of connecting the viewer
and the object, by stepping back from the position
of protagonist. And it is through intricate detail
that it manages not to lose a sense of equilibrium
in its existence.

The architecture of IDR resembles this.
Rather than splendidly coming to the fore, it
becomes part of the background, connecting
the detached and making one discover what was

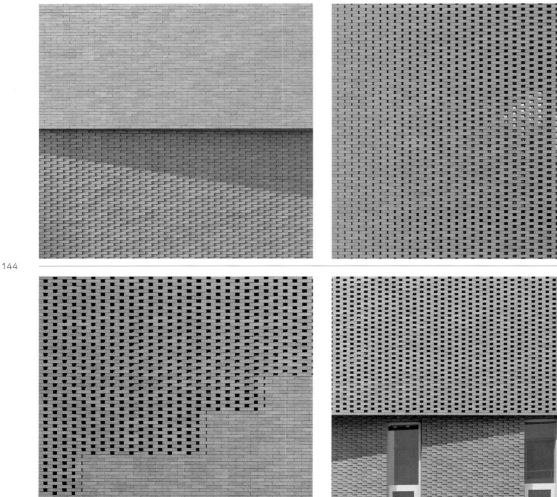

계양5동주민센터 계획안 (2016)
여러 재료를 사용하기보다는 하나의 재료를 다양한
방식의 디테일로 표현해 단순하면서도 단조롭지 않은 모습을
만들고자 했다.

THE COMPETITION ENTRY FOR GYEYANG
3-DONG COMMUNITY CENTER (2016)
The architects wanted to create a simple but
not monotonous look by constructing material
in various ways in detail.

몰랐던 것을 발견하게 하는 바탕의 건축, 그러면서 모종의
규칙을 통해 조심스럽게 자신의 존재를 드러내는 건축.

    이런 배경의 전략은 건축가가 프로젝트를 마주하는
과정에서 의식적으로 개입한다기보다는 건축가의 고유성이
무의식적인 차원에서 구체화하는 결과로 보는 것이 옳다.
간단히 말하면 집은 설계한 사람을 닮는 것이다. 우리의
작업을 뒤돌아보았을 때 이런 구체화는 서로 다르면서도
동시에 밀접하게 연관된 두 층위에서 일어나고 있음을
짐작할 수 있다.

    먼저 도시적 관점에서 살펴보자. 이런 접근은 마치
조각보와 같은 열악한 우리네 도시 환경에 조응하는
태도로 이해할 수 있다. 이는 본질적으로 주변 콘텍스트와
개별 건축의 관계를 설정하는 문제로, 단순히 콘텍스트를
참조하는 작업과는 차이가 있다. 도시 조직의 특징을
정의하는 조건을 찾아내 해석하고 이를 건축 행위로
재구축하는 과정을 따르는 작업인 것이다. 그 결과 도시라는
배경에 동화되면서 역설적으로 기존의 건축물과 일정한
거리를 유지하는 '낯섦'의 효과를 얻는다.

    인천 계양구의 ‹계양3동주민센터 계획안›은 치장 벽돌
마감의 다가구 주택이 들어찬 도시 조직과 접점을 찾기 위해

previously unseen by working as a setting, and and
at the same time carefully unveils its existence
through certain principles of its own. The strategy
of being a background is more of a result of
the architect's uniqueness being unconsciously
articulated, rather than intentionally intervening
in the process of developing a project. Put simply,
a house resembles the person who designed it. When
we look back on our projects, we can estimate that
such articulation is activated on two different yet
simultaneously intricately related levels.

    First, let's think of this from the city's
perspective. Such an approach could be understood
as a strategy to deal with our scrap-like urban
environments. This is fundamentally an issue of
setting the relationship between the surrounding
context, and the individual architectural project,
and hence differentiates from projects which simply
refer to the context. The conditions which define
the urban fabric are sought out and interpreted,
and an architectural project can be defined as a
process where these conditions are reconstructed.
As a result, this acquires a "defamiliarization"
which maintains a certain distance with the
existing buildings while ironically assimilating
with the city as a background.

    The competition entry for <Gyeyang 3-dong
Community Center> in Gyeyang-gu, Incheon was
based upon a strategically chosen method of using
moderately toned bricks on a simple yet intricately

차분한 색상의 벽돌을 단순하면서도 미묘하게 분할된 면에
여러 방식으로 쌓는 전략을 선택했다. 그 결과 그 동네에
속한 건축물처럼 보이면서도 전혀 다른 느낌으로 혼란스러운
주변을 환기하는 힘을 지니고 있다.

서울 명동 한복판에 자리 잡은 ‹명동 센트럴› 또한
주변 콘텍스트를 존중해 전벽돌과 저철분 유리로 차분하게
건물의 격을 드러낸다. 동시에 입점한 두 매장을 구분하기
위해 입면의 깊이와 패턴에 미묘한 차이를 주는 전략을
이용해 차별된 인상을 주었다. 주변의 다른 건물은 입점한
매장 인테리어가 외부까지 확장되어 보이지만, ‹명동
센트럴›은 멀리서는 하나로 보이면서도 다가갈수록 두 개의
면이 구분되는 이중적 태도를 취하기 때문에 주변과는 다른
지점에 있다.

‹풍무도서관 계획안›과 ‹호계문화체육센터 계획안›은
여러 면에서 비슷하다. 둘 다 녹지 언저리에 팽창 중인
주거 지역에 들어설 편의 시설로 계획되었다. 우리는 설계
공모 제출안에서 건물 크기를 차분하게 억제하면서 건축과
유기적으로 얽힌 조경을 최대한 확장해 공원이나 운동장과
같은 기존의 도시 맥락에 편입되고자 노력했다. 또한 기다란
입면을 수직적으로 분절해 스케일을 주변에 맞게 조절하되

divided surface to stack them in a variety of
different ways, as a way to find the intersection
with the urban fabric which was replete with
multi-family house built with ordinary bricks.
As a result, the project seems to be a building
originating from the town, yet carries a force
which can reventilate the chaotic surroundings with
an entirely different aura.

&lt;Myeongdong Central&gt; situated at the
center of Myeong-dong, Seoul, also exhibits its
sophistication through its calm presence using dark
bricks and ultra clear glass in respect to the
surrounding context. To distinguish the two stores
which share the same building, one section is
differentiated from the other by using the strategy
of slightly distinguishing the depths and patterns
on the façade. While the surrounding buildings seem
as if the interior of the stores have expanded
outwards, &lt;Myeongdong Central&gt; may be seen as one
building from a distance, but when one gets closer,
it shows a sense of duality by exposing the subtle
differences in the details between two surfaces,
thus distinguishing itself from the surrounding
environment.

The competition entries for &lt;Pungmu
Library&gt; and &lt;Hogye Cultural Sports Center&gt;
are similar in many aspects. Both were planned
as convenience facilities to be constructed in
expanding residential areas at the edge of green
buffer. In our plan, we strived to appropriate

명동 센트럴 (2019)
전벽돌과 저철분 유리로 만들어진 입면은 하나의 건물이지만
두 개의 매장이 나란히 들어서 있는 특수한 조건을 반영해
나온 것이다.

MYEONGDONG CENTRAL (2019)
The facade was designed to reflect the
special conditions of two stores side by side
in a building.

풍무도서관 계획안 (2016)
정면 투시도. 책을 꽂는 서가의 모습에서 모티브를
얻어 디자인했다.

THE COMPETITION ENTRY FOR
PUNGMU LIBRARY (2016)
Front perspective view. It is an elevation
designed with a motif from the bookshelf.

역시 기존의 건축물과는 사뭇 다른 표정을 지니는 결과가
되도록 의도했다. 최대한 화려해보이는 조형 어휘로 대지
전체를 장악하도록 설계된 당선작과는 정반대의 접근 방식인
것이다.

　　사용자 관점에서 보면 배경으로서의 건축은 실제로
사용되는 무대와 같다. 건축의 완성되는 시점은 준공
시점이 아니라 사람들이 건물과 관계를 맺고 건축가가 원래
의도했던 기능이 어느 정도 자리를 잡기 시작하는 때라는
이야기다. 물론 그 시점이 건물이 가장 아름다워보이는 때는
아닐 수도 있다. 그러나 일단 누군가의 사용을 가정하고
지어진 건물이라면 피할 수 없는 일이기도 하다.

　　〈매곡도서관〉 실시 설계 당시 내부가 너무 하얗고
밋밋하지 않냐고 지적하던 공무원을 설득하기 위해 우리가
펼쳤던 논리도 같은 맥락이다. 가구와 책이 가득 들어찼을
때를 생각하면 지금 정도의 인테리어만으로도 충분하다고
주장했다.

　　〈매곡도서관〉의 서쪽 입면 또한 원래는 그 앞에
계획되어 있던 정원의 배경 역할을 기대했다. 예산 문제로
조경이 대폭 축소되면서 그 의미는 많이 퇴색되었지만,
수직의 나무 무늬 루버는 실제 나무와 어우러지기를 바라는

the site in connection to the existing park and
sports facilities by assertively restricting the
size of the building, while maximally expanding
the landscape to be organically entwined with
the context. Also, we vertically partitioned the
long façade to modulate the scale to fit with its
surroundings while intending to look distinguished
from the existing buildings. Our proposal was
the very opposite of the winning one, which was
designed to dominate the entirety of the site with
very eye-catching form.

　　From the perspective of the user,
architecture as a background is actually like a
stage in use. This means that the moment of a
project's completion is not when the construction
has been finished, but when people start to form
relationships with the architecture and the
functions originally intended by the architect
start to settle in. Of course, that moment may not
be when the project looks its best. Nevertheless,
such a phase is unavoidable as a building is built
with the assumption that somebody will use it.

　　This is consistent with the logic we used to
persuade the public officials who alerted us during
the working design that the interior was too white
and bland for the <Maegok Library>. We asserted
that the interior was more than enough if one
were to consider the furniture and the books that
would fill the building. We had originally hoped
that the the facade on the west of the <Maegok

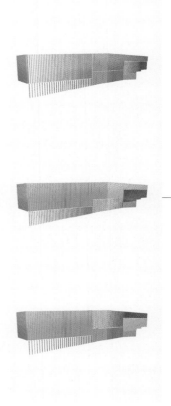

ↆ 〈매곡도서관〉 입면을 구성하는
루버는 패러매트릭 디자인 툴을
사용해 디자인되었다. 어떤 방향에서
보더라도 기분 좋은 리듬감을 느낄 수
있길 바랐다.

↗ The louvers that cover up
the <Maegok Library> were
designed using parametric
design tools. The architects
hoped that pedestrians could
feel a pleasant rhythm when
they see the building even
from various directions.

매곡도서관 (2017)
가구가 들어가기 전의 열람실. 바탕은 재료가 본래 지닌 색과
흰색만을 사용해 마감했다.

MAEGOK LIBRARY (2017)
The Reading room. The interior is finished
applying only natural materials and white.

매곡도서관 (2017)
매곡천과 마주한 도서관의 입면 모습. 숲을 이루는 나무의
수직적인 느낌을 형상화했다.

MAEGOK LIBRARY (2017)
The facade of the library facing Maegokcheon
River. It embodies the feeling of the trees
forming the forest.

마음에서 선택했다. 동시에 루버의 반복과 변화를 통해
적절히 통제된 복잡성이 현대적 도서관으로서의 성격을
암시할 수 있기를 바랐다.

　　이렇게 반복과 변화라는 주제를 갖고 배경의 조형을
정의하는 수법은 두 다목적강당에서도 마찬가지였다.
네 개의 동일한 트러스 truss 구조 모듈에서 출발한
〈언북중학교 다목적강당〉은 다소 엄격하고 큼직한 덩어리의
반복으로 건물 형태를 빚어내지만, 하늘과 땅의 연장을
암시하는 두 개의 수평 요소로 구성된 〈압구정초등학교
다목적강당〉은 접힌 종이 같은 면이 반복해 드러난다.
이는 〈풍무도서관 계획안〉과 〈호계문화센터 계획안〉에서
입면을 분할한 방식과 비슷하다. 두 곳 다 다양한 활동이
벌어지는 운동장에 나름의 방식으로 존재감 있는 배경으로
작용하고 있다.

　　우리는 건축이 배경으로서 물러났을 때 인간의
내면세계와 외부의 물리적 환경을 이어주는 매개체가 된다고
생각한다. 마치 피츠행어저택의 창문처럼. 배경이 된다는
것은 중요하지 않은 존재로 가라앉는다는 의미가 아니다.
주어진 조건을 해석하고 다시 구축하는 과정을 통해, 반복과
변화를 바탕으로 만들어내는 내재적 질서를 통해 그리고

Library> would play the role of a background for
the garden planned for the front of it. With the
landscape being minimized due to budget issues, the
significance of this has greatly diminished, yet
the vertical wooden patterned louvers were chosen
with the hope that it would harmonize with actual
trees. Simultaneously, we hoped the repetition
and modulation of the louvers would create an
appropriately restrained complexity to evoke its
character as a contemporary library.

　　This method of defining the form of the
background through the theme of repetition and
modulation also applied to the two multipurpose
auditoriums. While the four truss structure modules
which support <Eonbuk Middle School Multipurpose
Auditorium> molded the form of the building through
the repetition of a relatively large and somber
mass, the <Apgujong Elementary School Multipurpose
Auditorium> which was composed of two horizontal
elements to evoke the extension of the sky and
land, is similar to the method of dividing up the
façade of the competition entries for <Pungmu
Library> and <Hogye Cultural Sports Center>, giving
a sense of repeating the similar surfaces like
folded paper. Both are activated as backgrounds
with a sense of presence in their own ways within
the school yard where a myriad of activities take
place.

　　We believe that when architecture subsides
as a backdrop, it can become a medium which

언북중학교 다목적강당 (2018)
빛의 상자와 벽돌의 매스, 루버의 반복으로 생긴
리듬감 있는 입면. 네 개의 트러스 구조 모듈이 돋보인다.

EONBUK MIDDLE SCHOOL MULTIPURPOSE
AUDITORIUM (2018)
Elevation composed of a box of light,
a mass of bricks, and a louver. Four truss
structural modules stand out.

압구정초등학교 다목적강당 (2018)
종이접기처럼 다양한 각도로 접힌 벽면이 특징이다.
재료는 운동장의 색을 느낄 수 있는 벽돌로 선택했다.

APGUJONG ELEMENTARY SCHOOL MULTIPURPOSE
AUDITORIUM (2018)
It is a wall folded at various angles like
origami. The wall material was chosen as
brick similar to the color of the playground.

무엇보다 드러내고 물러나는 것의 적절한 균형점을 찾는
과정을 통해 배경은 그 안에서 이루어지는 경험의 텍스처를
결정하기 때문이다. 그렇게 우리의 삶이 외부 세계가 가진
풍요로움과 좀 더 특별하고 윤택한 관계에 놓이면 좋겠다.
그것이 빛이든 어둠이든 바람이든 풍경이든 도시든
그 무엇이든 간에.

connects our inner world with the physical
environment around us, like the window from the
Pitzhanger Manor House. Becoming the background
does not mean that its sinks to the level of
unimportance. Through the process of interpreting
and reconstructing the existing conditions, and
the inherent order established by the grammar
of repetition and modulation, and above all
the elaboration on seeking out an appropriate
equilibrium between emerging and subsiding,
the background determines the textures of the
experience which are to occur within. As such, we
hope that our life might have a more special and
plentiful relationship with the richness of the
outer world, whether it be light, darkness, wind,
the scenery, the city, or whatever it may be.

# 투쟁

# Struggle

Essay Five

#설계비 #매곡도서관 #압구정초등학교 #언북중학교
#다목적강당 #학교건축 #디테일 #전문가

이것 한 가지는 분명히 해두고 싶다. 아이디알은 절대로
호전적 성격의 사람들이 아니다. 이승환은 주변의 모든
사람과 원만한 관계를 맺으려고 하는 성격이고 전보림은
예의 바르게 행동하려고 꽤 신경을 쓰며 산다.

　　그러나 불의나 불합리를 보면 못 참고 욱하는 성격이
살짝 있기는 하다. 특히 대외적 소통을 담당하는 전보림이
조금 더 그렇다. 그래서 다른 사람, 즉 다른 건축가 같으면
그냥 어쩔 수 없는 것으로 생각하고 넘어갔을 일을 '아니
대체 왜?' '이건 정말 말도 안 되는 일이야.'라고 생각한다.
그리고는 "그래도 한 번 이야기는 해봐야지." "어쨌든 우리가
할 수 있는 일은 다 해봐야지." 하면서 그걸 정말로 실천에
옮기고야 만다.

　　시작은 ‹매곡도서관› 계약이었다. 발주처에서 설계
공모 때 공고한 설계비 그대로 계약하는 것이 아니라 금액을
깎아서 계약하겠다고 했다. '아니, 대체 왜?' 이런 의문이
들어 물었더니 관례라는 대답이 돌아왔다. 다른 프로젝트도
다 그렇게 계약을 했기 때문에 이 프로젝트도 그렇게 해야만

#DesignFees #MaegokLibrary
#ApgujongElementarySchool #EonbukMiddleSchool
#MultipurposeAuditorium #SchoolArchitecture
#Detail #Expert

We would like to make one thing clear. We are
not the people who like to fight on everything.
Seunghwan Lee tries to have good relationships with
everyone around him, and Borim Jun makes an effort
to be polite.

　　Nevertheless, we are triggered when
witnessing injustice or illogical things. This is
slightly more pronounced with Borim Jun who takes
on the public relations role of the firm. She often
thinks "why in the world?" or "this doesn't make
any sense at all" to things that others, or other
architects might accept and think of as "things
that are inevitably a certain way." She then goes
on to think "we should still try to discuss this" "we
should still try everything we can" and actually
goes ahead to realize this.

　　This started with the contract for the
<Maegok Library>. The commissioning body declared
that they would sign the contract at a rate lower
than the design fee they had notified during the
competition. The response she received after
inquiring "Why in the world?" was that it was
just the custom. The logic was that all the
other previous projects had been contracted as
such, and hence it should follow suit. They said

한다는 논리였다. 안 그러면 감사를 받는다고 했다. 여기저기 전화해서 물어보고 방법을 알아보았다. 멀리 경상북도 영주시 수영장 설계 계약 당시 설계비를 깎지 않고 계약한 사례가 있어 찾아서 보여주었지만, 울산시에 아무런 영향을 주지는 못했다. 지금의 지방자치단체 시스템에서 다른 시는 다른 나라나 마찬가지였다. 하는 수 없이 전보림이 발표 자료를 만들어 울산으로 갔다. 젖먹이 셋째를 대동하고. 요지는 이러했다.

> "구청장님이 말씀하신 명품 건축이 되기 위해서는 디테일이 중요합니다. 그 디테일을 제대로 만들기 위해서는 도면을 많이 그려야 하죠. 그것이 고급 설계입니다. 우리는 도면을 많이 그리는 고급 설계를 하니 설계비를 제대로(깎지 말고) 받아야 합니다."

8개월 아기를 혼자 데리고 서울에서 울산까지 달려온 아줌마 건축사의 발표를 재무과 주무관과 담당 건축과 주무관들은 무덤덤한 표정으로 들었다. 대답은 물론 '안 된다'였다.

그때만 해도 몰랐다. 공모에 당선된 뒤 설계비를 감액하는 관행인 수의시담隨意示談도 몰랐고 공공 건축물

that there would be auditing if they didn't. She started making phone calls, looking for any other possibilities. She looked up a case where a public contract for designing a swimming pool in Yeongju-si, Gyeongsangbuk-do far away had gone ahead without deducting the fee, but this had no impact on what went on in Ulsan. In the current system of autonomous local governments, different cities are comparable to foreign countries. With nothing more she could do Borim Jun prepared a presentation and went to Ulsan, accompanied by her breastfeeding third child. The point of the presentation was,

> "Detail is paramount if we want to realize the quality architecture cited by the district mayor. To make sure those details are made properly, we need to produce lots of drawings. That is what we call high quality design. We achieve this by producing more drawings so we need the total design fee (without a deduction)."

The public officials from the financing department and the managing architecture department listened without a hint of expression to the presentation by this fussy lady who had travelled alone from Seoul to Ulsan with her eight months old baby. Of course, their response was a resounding "no."

Up until that point we had no idea. We had no knowledge of bidding settlements, the custom where

매곡도서관 (2017)
(좌) 실내 계단 디테일.
(우) 강당으로 가는 복도의 천창.

MAEGOK LIBRARY (2017)
(left) The details of the interior staircase.
(right) The skylight of hallway towards
auditorium.

공사를 운영하는 공무원의 세계도 몰랐다. 공무원이 기존의
틀에서 벗어나 다르게 일하기가 얼마나 어려운지, 아니
사실 거의 불가능한 일에 가깝다는 것을. 설사 그 틀이 매우
불합리할지라도 말이다. 하지만 전보림은 생각했다. 바꿀 수
없다고 해서 불합리함을 그대로 수긍하는 것은 그에 대한
암묵적 동의 아닐까? 설사 우리가 하는 항의가 소용없더라도
적어도 어떤 부분이 불합리한지는 알릴 필요가 있지 않을까?
지금은 모르겠지만, 언젠가 이들도 그때 그 아줌마 건축사가
했던 이야기를 이해하게 될 날이 올 것이다. ⟨매곡도서관⟩
설계는 다른 건물하고 다를 테고 그 결과가 설계의 차이를
보여줄 테니까.

　　　계약부터 만만치 않았으니 뭐 하나 쉽지 않았을
것으로 예상하겠지만, 사실 우리는 이후 납품에 이르기까지
비교적 순탄하게 일을 했다. 잘 몰라서 헤맨 적도 없지는
않았지만 담당 공무원들과 매우 안정적인 신뢰 관계를
유지했다. 납품하면서 디자인 감리를 할 수 있게 해달라고
간곡하게 요청했으나 거절당한 것, 시공 과정에서 일부
재료를 변경하는 것에 대해 원안 유지를 부탁한 것 정도의
자잘한 사건만 있었을 뿐이다. 비교적 설계자를 존중하는
발주처였고 다행히 설계안의 가치를 알아보는 담당자를

the design fee is reduced after being nominated
for the competition, and we also had no idea about
the world of public officials who manage these
public institutions. We also had no idea about how
difficult it was for public officials to deviate
from existing norms to work a different way, and
that this was practically impossible, even if this
framework is apparently unjust. Borim Jun would
think, if you accept injustice as it is, just
because you can't change it, isn't this some kind
of implicit agreement? Even if it was useless to
complain, wasn't it at the very least, necessary
to make it known how unjust this was? While they
might not know it now, the day will come when they
will understand what that lady architect was going
on about. The design of the <Maegok Library> would
be different from that of other public projects,
and the result will prove that the difference comes
from the design.
　　　One might think that nothing would have gone
smoothly, with the difficulties of the contract
from the start. Nevertheless, the work flowed quite
smoothly after this, until the final delivery.
There were times when we got lost in the situation,
yet we maintained an extremely stable relationship
of trust with the project officers. There were
only a couple of hiccups, such as when our earnest
request was rejected to supervise the construction,
or when we had to ask them not to change materials
for some interior walls during construction.

만났으며 감사하게도 원리원칙을 지켜 도면 그대로
시공하도록 지시한 감리자 덕분에 비교적 원래의 설계대로
지어질 수 있었다.

　　그러나 그다음 프로젝트인 〈압구정초등학교
다목적강당〉과 〈언북중학교 다목적강당〉에서는 그야말로
본격적인 투쟁의 연속이었다. 계약 때 수의시담을 들먹이며
공고된 설계비의 고작 87퍼센트 금액으로 계약하겠다고 해서
항의했고, 교장 선생님과 가진 첫 회의에서 담당 주무관이
"〈압구정초등학교 다목적강당〉의 접힌 입면을 나중에 설계
변경으로 쫙 펴버리겠다."라고 해서 속으로 기함하기도
했다. 건물에 사용할 제품과 재료를 지정하려 하니 교육청
주무관들이 저지해서 처음에는 설득으로, 나중에는 언쟁으로
발전하기도 했다. 벽돌 색깔과 타일은 건물의 주인인
학교장이 결정하는 것이라는 논리 앞에서 순간 어이가 없고
기가 막혀 말이 안 나왔지만, 그렇다고 아무 말도 안 하고
넘어가서는 안 되는 상황이었다.

　　그렇다. 가장 분노했던 몇몇 순간은 설계자인
건축가를 전문가로서 존중하는 것은 고사하고 인정조차 하지
않는 상황이었다. '건축가님'이라고 떠받들어주기를 기대한
건 절대 아니었다. 그저 최소한 건축 디자인이 전문가의

The building was constructed to be relatively
consistent with the original design due to a
commissioning body who was relatively respectful
of the architect, and a manager who fortunately
recognized the value of the design, and an
supervisor who thankfully directed the construction
to be built in line with the principles of the
drawings.
　　However, the following two projects,
<Apgujong Elementary School Multipurpose
Auditorium> and <Eonbuk Middle school Multipurpose
Auditorium>  were indeed a continuing struggle.
We protested against their bid to conclude the
contract at just 87% of the sum originally outlined
in the competition, based on the referenced bidding
settlements during the contract process, and we
were struck dumb at the words of the project
officer in the first meeting with the school
principal, who said "we'll just flatten out the
folded bits on façade when we eventually modify
the design" referring to the <Apgujong Elementary
School Multipurpose Auditorium>. While designating
the products and materials for the building,
we countered the skepticism of the Office of
Education's officers at first by coaxing them, and
then by quarreling. We were left lost for words the
moment we confronted the logic that the color of
the bricks and tiles should be determined by the
school principals as the "owner" of the school,
yet, it wasn't a situation where we could stand

압구정초등학교 다목적강당 (2018)
미세하게 서로 다른 각도로 접혀 있는 방식의 입면이라
햇빛을 받으면 각자 다른 색과 질감을 띤다.

APGUJONG ELEMENTARY SCHOOL MULTIPURPOSE
AUDITORIUM (2018)
Façades are folded at slightly different
angles, so they show different colors and
textures when exposed to sunlight.

다목적강당 인테리어
(위) 압구정초등학교 다목적강당 계단실.
(아래) 언북중학교 다목적강당 1층 로비.

INTERIOR VIEW OF THE AUDITORIUMS
(top) The staircase hall at Apgujong
Elementary School.
(bottom) The entrance hall at Eonbuk Middle
School.

영역이며 전문가의 선택을 존중하면 더 나은 결과를 만들
수 있다는 인식 정도만이라도 있을 것이라 생각했다. 그런데
존중은 고사하고 학교 건축의 진행 과정에서 건축가는
전문가로 인정받지도 않는 것 같았다. 어떤 주무관은 자신이
다목적강당을 벌써 몇십 개를 한 전문가라고 말했고, 또
어떤 주무관은 우리가 설계한 다목적강당 1층의 급식 시설
설계를 전자 입찰로 발주해버리고는 자신은 모든 건축사의
능력을 믿는다며 누가 설계하든지 좋은 결과가 나올 거라고
했다. 한 학교 관계자는 시공까지 끝난 건물의 벽면 디자인을
바꿔주길 요구했고 어떤 학부모는 자기가 색채 전문가이니
건물에 칠한 색상을 바꿔달라고 했다. 어떻게 그 앞에서
아무 말도 하지 않고 아무 감정도 드러내지 않고, '네' 하고
수긍하며 순순히 하라는 대로만 할 수 있겠는가? 도저히 그럴
수 없다. 싸우는 것을 좋아해서 싸운 것이 결코 아니었다.

　　우리가 맞닥뜨린 상황이 학교 건축만의 이야기는 아닐
것이다. 어쩌면 건축 디자인 즉 건축 설계에 대한 이 사회의
인식 수준이 얼마나 낮은지 그 민낯이 드러난 것뿐일지도
모르겠다. 안전에 관한 것만 아니면 너도 나도 이래라저래라
할 수 있고 마음대로 바꿔도 된다는 생각. 설계는 최대한
저렴할수록 좋다는 생각. 주인이 재료를 선택해야 한다는

aside and say nothing.
　　Yes. The moments we felt most enraged was
when there was no recognition, let alone respect
for the designing architect as an expert. We had
by no means hoped that they would put us on a
pedestal, calling us "the architect." We just
hoped that there would be a glimmer of recognition
that at the least, architectural design was an
expert field, and that reverence for the choice
of the expert would lead to better results. Yet
the process of the school architecture made us
feel that architects were not even recognized as
professionals, let alone respected. One officer
told us that he was an expert, having made a
couple dozen multipurpose auditoriums, and another
officer, who commissioned the cafeteria design on
the first floor for the multipurpose auditorium we
had designed as digital bidding, then declared that
she believed in the abilities of all registered
architects, and that good results would come from
whoever the designer might be. One school official
requested that the wall design of the building
which had already completed be modified, and a
parent of a student asked us to change the color
of the building since he was a color expert. How
could we remain expressionless, say nothing, accept
everything and do whatever they told us to? There
was no way we could. It wasn't because we enjoy
fighting that we fought so hard.
　　The situation we faced does not solely apply

연북중학교 다목적강당 (2018)
내부 시트지가 시공되기 이전 모습.

EONBUK MIDDLE SCHOOL MULTIPURPOSE
AUDITORIUM (2018)
The auditorium interior at Eonbuk Middle
School. Before the setting the black sheets.

생각. 미술 공부한 사람이 색상을 더 잘 고를 것이라는 생각.
세금으로 지은 공공 건축이어도 사용자가 곧 주인이니
사용하면서 마음대로 바꿔 써도 된다는 생각.

　　동의할 수 없는 상황을 만날 때마다 화를 내며 싸우는
것은 아니다. 사실 분노와 언쟁만으로 세상을 바꿀 수 없다는
사실을 우리도 안다. 그러나 아름다운 건축은 세상을 바꿀
힘이 있다고 믿기에 좋은 건축을 만들기 위해 노력하는
것뿐이다. 그 과정에서 불합리한 일을 만나면 때로는
목소리를 높여 언쟁하고 공문을 보내기도 하며 매체에
울분의 글을 쓰기도 한다. 도면을 열심히 그리는 것만으로
충분치 않을 때만 그렇게 하는 것이다. 그런 모습에
'투쟁'이라는 이름을 붙여야 하는 것이 우리도 싫다. 우리는
매우 예의 바른 사람들이기 때문에.

↖ 〈언북중학교 다목적강당〉 강당 내부.
빛의 띠로 계획된 창에는 블라인드가
설치될 예정이었으나 사용자가 임의로
시트지를 부착했다.

169

to the field of school architecture. Perhaps this
was a moment which exposed the bare face of what
little awareness there is in this society about
architectural design. The common opinion is that
it would be ok to change whatever, by whoever, as
long as it has issues with safety, or that the
cheaper, the better the design, or that the owner
should be allowed to choose the materials, or that
someone who studied art would be better at choosing
color schemes, or even that since the users are the
owners of buildings they can change whatever they
want, even for the public facilities built with
taxpayer's money.

　　We don't get angry and start fighting every
time we confront a disagreeable situation. We
understand that one can't change the world by
being angry and quarreling with others. Yet, we do
believe that great architecture has the capacity
to change the world, and so we have to try to make
good architecture. When we come across something
unjust in that process, at times, we do have to
raise our voices and fight, or send off official
complaints, or write to the media about our woes.
We only do this when drawings we make turns out
not to be quite enough. We also feel uncomfortable
about the fact that this process must inevitably be
labeled as a "struggle." We are very polite people.

↗ Interior view of the
<Eonbuk Middle School
Multipurpose Auditorium>.
It was a window for the
skylight, but the user stuck
the black sheets and blocked
the lights.

설계 개요

매곡도서관

· 설계 ¦ 아이디알 (전보림, 이승환)
· 위치 ¦ 울산광역시 북구 매곡로 138-19
· 용도 ¦ 교육연구시설(공공도서관)
· 대지면적 ¦ 3,787m²
· 건축면적 ¦ 731.34m²
· 연면적 ¦ 2,103.77m²
· 규모 ¦ 지상 3층, 지하 1층
· 높이 ¦ 12.75m
· 주차 ¦ 20대
· 건폐율 ¦ 19.31%
· 용적률 ¦ 43.8%
· 구조 ¦ 철근콘크리트조
· 외부마감 ¦ 알루미늄 루버, 송판무늬 노출콘크리트, 외단열 미장 마감
· 내부마감 ¦ 콘크리트 면처리 위 수성페인트, 알루미늄 루버 천장
· 구조설계 ¦ 하모니구조
· 기계·전기설계 ¦ 하나기연
· 시공 ¦ 미건종합건설
· 설계기간 ¦ 2015. 8. - 2016. 2.
· 시공기간 ¦ 2016. 4. - 2017. 4.
· 사진 ¦ 전영호
· 건축주 ¦ 울산광역시 북구

Design overview

MAEGOK LIBRARY

· Architect ¦ IDR (Borim Jun, Seunghwan Lee)
· Location ¦ 138-19, Maegok-ro, Buk-gu, Ulsan, Korea
· Programme ¦ educational facility (public library)
· Site area ¦ 3,787m²
· Building area ¦ 731.34m²
· Gross floor area ¦ 2,103.77m²
· Building scope ¦ B1, 3F
· Height ¦ 12.75m
· Parking capacity ¦ 20
· Building coverage ¦ 19.31%
· Floor area ratio ¦ 43.8%
· Structure ¦ RC
· Exterior finishing ¦ aluminium louver, board formed concrete, exterior insulation finishing system
· Interior finishing ¦ water based paint on concrete, aluminium louver ceiling
· Structural engineer ¦ Harmony Structural Engineering
· Mechanical and electrical engineer ¦ HANA Consulting Engineers Co., LTD.
· Construction ¦ Migun Construction
· Design period ¦ Aug. 2015 - Feb. 2016
· Construction period ¦ Apr. 2016 - Apr. 2017
· Photograph ¦ Youngho Chun
· Client ¦ Buk-gu Office, Ulsan

언북중학교 다목적강당
- 설계 ¦ 아이디알 (전보림, 이승환)
- 위치 ¦ 서울시 강남구 도산대로38길 27
- 용도 ¦ 교육연구시설(다목적강당)
- 대지면적 ¦ 13,197m²
- 건축면적 ¦ 879.61m² (증축 부분)
- 연면적 ¦ 993.41m² (증축 부분)
- 규모 ¦ 지상 2층
- 높이 ¦ 16.75m
- 건폐율 ¦ 25.35% (전체)
- 용적률 ¦ 74.77% (전체)
- 구조 ¦ 철근콘크리트조, 철골조(지붕)
- 외부마감 ¦ 적벽돌, 폴리카보네이트 패널, 칼라 강판
- 내부마감 ¦ 콘크리트 면처리 위 수성페인트,
시멘트 블럭 위 수성페인트, 흡음보드, 고무안전리브
- 구조설계 ¦ 포은구조
- 기계·전기설계 ¦ 하나기연
- 시공 ¦ 근복공영
- 설계기간 ¦ 2017. 1. – 4.
- 시공기간 ¦ 2017. 10. – 2018. 10.
- 사진 ¦ 노경
- 건축주 ¦ 서울특별시 교육청

압구정초등학교 다목적강당
- 설계 ¦ 아이디알 (전보림, 이승환)
- 위치 ¦ 서울시 강남구 압구정로39길 29
- 용도 ¦ 교육연구시설(다목적강당)
- 대지면적 ¦ 14,135.6m²
- 건축면적 ¦ 837.83m² (증축 부분)
- 연면적 ¦ 943.74m² (증축 부분)
- 규모 ¦ 지상 3층
- 높이 ¦ 13.7m
- 건폐율 ¦ 18.75% (전체)
- 용적률 ¦ 51.75% (전체)
- 구조 ¦ 철근콘크리트조, 철골조(지붕)
- 외부마감 ¦ 콘크리트 벽돌, 폴리카보네이트 패널
- 내부마감 ¦ 콘크리트 면처리 위 수성페인트, 흡음보드, 고무안전리브
- 구조설계 ¦ 포은구조
- 기계·전기설계 ¦ 하나기연
- 시공 ¦ 우리글로벌건설
- 설계기간 ¦ 2017. 1. – 4.
- 시공기간 ¦ 2017. 7. – 2018. 11.
- 사진 ¦ 노경
- 건축주 ¦ 서울특별시 교육청

EONBUK MIDDLE SCHOOL MULTIPURPOSE AUDITORIUM
- Architect ¦ IDR (Borim Jun, Seunghwan Lee)
- Location ¦ 27, Dosan-daero 38gil, Gangnam-gu, Seoul, Korea
- Programme ¦ educational facility (multipurpose auditorium)
- Site area ¦ 13,197m²
- Building area ¦ 879.61m² (extended area)
- Gross floor area ¦ 993.41m² (extended area)
- Building scope ¦ 2F
- Height ¦ 16.75m
- Building coverage ¦ 25.35% (in total)
- Floor area ratio ¦ 74.77% (in total)
- Structure ¦ RC, steel (roof)
- Exterior finishing ¦ red brick, polycarbonate panel, color coated steel sheet
- Interior finishing ¦ water based paint on concrete, water based paint on cement block, acoustic board, rubber wall padding
- Structural engineer ¦ Poeun Structure
- Mechanical and electrical engineer ¦ HANA Consulting Engineers Co., LTD.
- Construction ¦ Keunbok Construction
- Design period ¦ Jan. – Apr. 2017
- Construction period ¦ Oct. 2017 – Oct. 2018
- Photograph ¦ Kyung Roh
- Client ¦ Seoul Metropolitan Office of Education

APGUJONG ELEMENTARY SCHOOL MULTIPURPOSE AUDITORIUM
- Architect ¦ IDR (Borim Jun, Seunghwan Lee)
- Location ¦ 29, Apgujeong-ro 39gil, Gangnam-gu, Seoul, Korea
- Programme ¦ educational facility (multipurpose auditorium)
- Site area ¦ 14,135.6m²
- Building area ¦ 837.83m² (extended area)
- Gross floor area ¦ 943.74m² (extended area)
- Building scope ¦ 3F
- Height ¦ 13.7m
- Building coverage ¦ 18.75% (in total)
- Floor area ratio ¦ 51.75% (in total)
- Structure ¦ RC, steel (roof)
- Exterior finishing ¦ concrete brick, polycarbonate panel
- Interior finishing ¦ water based paint on concrete, acoustic board, rubber wall padding
- Structural engineer ¦ Poeun Structure
- Mechanical and electrical engineer ¦ HANA Consulting Engineers Co., LTD.
- Construction ¦ Woori Global Construction
- Design period ¦ Jan. – Apr. 2017
- Construction period ¦ July 2017 – Nov. 2018
- Photograph ¦ Kyung Roh
- Client ¦ Seoul Metropolitan Office of Education

명동 센트럴
· 설계 ┆ 아이디알 (전보림, 이승환)
· 위치 ┆ 서울특별시 중구 명동8길 17-1
· 용도 ┆ 근린생활시설(판매시설)
· 대지면적 ┆ 381.4m² (증축 부분)
· 건축면적 ┆ 330.8m² (증축 부분)
· 연면적 ┆ 926.2m²
· 규모 ┆ 지상 4층
· 높이 ┆ 19.6m
· 주차 ┆ 0대(주차제한구역)
· 건폐율 ┆ 86.73% (전체)
· 용적률 ┆ 242.84% (전체)
· 구조 ┆ 철골조
· 외부마감 ┆ 전벽돌, 투명유리, U자형 글라스
· 구조설계 ┆ 사림엔지니어링
· 기계·전기설계 ┆ 하나기연
· 시공 ┆ 제이케이에스 건설
· 설계기간 ┆ 2018. 1. - 6.
· 시공기간 ┆ 2018. 6. - 2019. 1.
· 사진 ┆ 노경
· 건축주 ┆ 박지호 외 12인

MYEONGDONG CENTRAL
· Architect ┆ IDR (Borim Jun, Seunghwan Lee)
· Location ┆ 17-1, Myeongdong 8-gil, Jung-gu, Seoul, Korea
· Programme ┆ retail
· Site area ┆ 381.4m²
· Building area ┆ 330.8m² (extended area)
· Gross floor area ┆ 926.2m² (extended area)
· Building scope ┆ 4F
· Height ┆ 19.6m
· Parking capacity ┆ 0
· Building coverage ┆ 86.73% (in total)
· Floor area ratio ┆ 242.84% (in total)
· Structure ┆ steel
· Exterior finishing ┆ clay brick, clear glass, U-shaped glass
· Structural engineer ┆ Salim Engineering
· Mechanical and electrical engineer ┆ HANA Consulting Engineers Co., LTD.
· Construction ┆ JKS Construction
· Design period ┆ Jan. - June 2018
· Construction period ┆ June 2018 - Jan. 2019
· Photograph ┆ Kyung Roh
· Client ┆ Jiho Park and 12 others

젊지 않은
: 김재관

Critique

The Not-So-Young
: Jaekwan Kim

김재관
충북 옥천 사람이다. 1997년에
무회건축연구소를 세운 후 개신교
교회와 주거 시설을 설계했으며 서울
문화의 밤 행사인 '일일설계사무소'에서
만난 ‹율리아의 집›을 수리하면서
집수리업자로 전향한 뒤 10년
동안 열다섯 채의 집을 수리했다.
한국건축문화대상, 경기도 건축상,
서울시 건축상을 수상했다. 지은
책으로는 『수리수리 집수리』가 있다.

젊은 건축가 — 그들

아이디알을 처음 알게 된 건 그들이
페이스북에 쓴 글을 통해서다.

> "경계없는작업실은 '현실의
> 건축'이라는 주제 하에 상업적 가치를
> 최대로 끌어올리면서도 이를 높은
> 건축적 완성도를 가진 결과물로
> 만들어내기 위해 떠올린 아이디어를
> 중심으로 이야기를 풀어나갔다.
> 그러한 방법들이 감탄을 자아낼 만큼
> 스마트했고, 적당한 선에서 멈추지
> 않고 두 마리 토끼를 다 잡기 위해
> 들인 노력의 지극함이 돋보였다.
> 푸하하하프렌즈는……(중략). 올해는
> 서른한 팀이 1차 포트폴리오를
> 제출했고 이중 일곱 팀이 선정되어
> 2차 공개 심사에 참여했다. 우리의
> 전략은 다음과 같았다."[1]

Jaekwan Kim
Born in Moohoi Village, South
Korea. After opening Moohoi
Architecture Studio in 1997,
he has designed protestant
churches and residential
facilities, then switched
the course of his work to
restoring houses after having
repaired ‹Julia's House›
whom he met at the 'Daily
Design Studio,' an event
of Seoul Open Night. He has
now repaired fifteen houses
over the past decade. He
has been awarded the Korean
Architecture Award, Gyeonggi
Province Architecture Award,
Seoul Architecture Award.
He has written 『Soori Soori,
Jip-soori』.

## YOUNG ARCHITECTS - THEM

I first came across IDR Architects
through something they wrote.

> "BOUNDLESS developed their
> presentation on the theme
> "architecture in reality" based
> on their idea to simultaneously
> maximize commercial value while
> augmenting the quality of the
> architecture. The methods they
> used were awe-inspiringly
> smart, accentuated by their
> tireless efforts to catch two
> birds with one stone instead of
> stopping at a certain limit.
> FHHH friends---- (omission). This
> year 31 teams submitted their
> portfolios for the first round,
> and among these, seven teams
> were designated to take part in
> the second round adjudication.
> Our strategy was as follows."[1]

The article was something I had
discovered on Facebook, with the
title "Participating at the 2018
Korean Young Architect Award
Competition," and recorded their

「2018년 젊은건축가상 참여기」라는 제목으로, 경합에서 탈락한 실패의 기록이었다. 왜 이런 글을 쓸까 싶었고, 분명 '짰다' '억울하다' '제도에 문제가 있다' 하는 말이 등장하겠거니 짐작했지만, 그런 종류의 글과는 달랐다. 물론 분한 마음이 전혀 없지는 않았겠지만(글과 함께 실린 사진에서 이승환의 물기 어린 눈빛과 전보림의 푸른 콧김을 보면), 분노보다는 서운함 정도였다. 그 순한 객관화들, 모진 타자화들 그리고 덤덤한 분석은 과하지 않았고 또한 의젓했다.

그러다 〈압구정초등학교 다목적강당〉 필화 사건이 터졌다. 발단은 학교 측에서 안전을 빌미로 이미 시공된 벽을 자기네가 정한 모양으로 다시 시공해 줄 것과 벽에 칠한 색상을 그들이 원하는 것으로 바꿔달라고 교육청에 요구한 공문 때문이었다. 사실 공공 건축에서 전문가인 설계자를 무시하고 사용자가 주인인 양

설계를 바꿔서 시공하는 일은 오래전부터 있어온 일이다. 그래도 대개는 푸념이나 수군댐으로 끝내는데 전보림은 「우리나라 학교 건축이 후진 진짜 이유」라는 글을 페이스북과 블로그에 올렸다. 거기에는 박탈된 재료 선택 권한을 비롯해 그동안 회자되던 발주처의 부당함이 낱낱이 적혀 있었다.

"일단, 교육청 시설 주무관들은 설계자가 재료를 지정하는 것을 이해하지도 못할뿐더러 대놓고 막는다. 재료는 학교장이 미술 선생님과 함께 선정하는 것이라고 하는데 사실은 시공 과정에서 감독관인 시설 주무관들이 개입하기 위한 것이다. 자기와 연결고리가 있는 업체의 샘플을 몇 개 학교에 가지고 가서 그 중에서 학교장이 고르도록 한다. 그러면 겉으로 보기에는

failure in the competition. I assumed the article would be sprinkled with expressions like "pre-determined" "resentful" "systemic problems." However, the article turned out to be quite different from anything that I had expected. Of course, this doesn't mean to say that they weren't vexed at all, (the photo depicts a misty-eyed Seunghwan Lee and a cloud of bluish haze around Borim Jun), yet the article was more than an expression of tearful injury or rage. The attempts to objectify the situation, the pointed effort to put it in perspective, and the composed analysis was not one of emotional excess, but rather of maturity.

Then, came the post about the ‹Apgujong Elementary School Multipurpose Auditorium›. It all started when school requested the Office of Education for the permission to reform a wall for safety issues, and repaint a wall with the color they wanted. Public architecture has long suffered from abuses of authority in revising the design without the consent from the architect. While most cases

are closed with nothing more than disgruntled moaning or gossip, Borim Jun had published an article on their blog and Facebook titled 'The real reason school architecture in Korea sucks.' The article listed in detail the architect's right to choose selected materials, as well as the aforementioned injustice of the commissioning body.

"First, the facility department officials of the Office of Education not only have no understanding of how the designer selected the materials, they go a step further in outright preventing this from happening. They say that materials should be designated by the principal of the school with the art teachers, but this is in fact to make way for the facility department officials who audit the construction process to intervene. They take a few samples from businesses they have ties with to the school and request that the principal choose from among this

학교에서 결정한 것이 되지만 사실은 교육청 공무원이 업체를 결정하는 것이다. 재료와 제품을 구체적으로 지정하지 못하도록 한 제도는 그간 교육청 공사에 만연한 청탁과 부패(그 유명한 교육청 카르텔)를 막기 위해 만들어진 것인데, 실제로는 이런 식으로 악용되고 있다. 나는 주무관이 싫어하든 말든, 설계 과정에서 미리 재료와 제품을 결정하겠다고 우겼다."[2]

소주잔을 기울이며 해소하던 어둠의 이야기가 드디어 세상에 알려졌다. 소셜 네트워크 서비스를 타고 급속히 퍼진 글은 결국 해당 공무원에게 알려지면서 법적 분쟁까지 언급되는 지경에 이르렀고 국가 건축 위원회에까지 알려지고 나서야 수습되었다. 사고를 크게 친 것이다. 결과적으로 이 사건은 공공 건축 프로젝트에 적지 않은 영향을 남겼는데 대표적인 것이

재료 선정에 대한 공무원의 불개입이다. 이후 공공 건축 분야에 진입한 건축가들은 한결 수월한 환경에서 일할 수 있게 되었다.

"아이디알 건축사사무소가 2019년 젊은건축가상 수상자로 선정되었습니다. 몹시 기쁘면서도 한편으로는 맘 편하게 마냥 기뻐하지만도 못하겠습니다. 아마도 작년에 2차 심사까지 받고서 탈락한 후에 느꼈던 진한 실망감이 떠올라서인 것 같습니다. 부끄럽게도 지난 일 년 동안 저희 스스로의 작업과 방향에 대한 의구심을 떨치지 못하고 살았습니다."[3]

「2018년 젊은건축가상 참여기」 이후 1년 만에 블로그에 쓴 「2019 젊은건축가상 수상」 글이다. 와신상담臥薪嘗膽, 권토중래捲土重來.

selection. This way, while it seems like it's the school who has made the decision, but it is actually the public officials at the Office of Education who are choosing the businesses. The system of not allowing specific designation of materials or the products was made to prevent the past widespread asking for favors and corruption (the notorious Office of Education cartel) in construction projects led by the Office of Education, but such practices were actually taking advantage of this. I insisted that whether the officials like it or not, we would decide the materials and products in advance in the design process."[2]

This dark tale, previously dealt with through bitter conversations and shots of soju, was finally made known to the world. The article, which rapidly went viral on social media became known to the public official in question, coming to a head where legal battles were proposed, and was

finally dealt with after it was made known to the National Architecture Commission. They had struck big. As a result, the incident had quite an impact on public projects, in particualr, the non-intervention of public officials in selecting materials. Through this, architects who have entered into the world of public architecture can now work in relatively less interference.[3]

"IDR Architects have been nominated as winners of the 2019 Young Architect Award. We are extremely happy, yet on the other hand, we cannot just be complacent. This is probably because it has reminded us of the deep sense of disappointment after our failure to win after the second round of evaluations last year. Embarrassingly, we have lived for the past year unable to cast off a sense of doubt about our own work and future direction."

This is the post "Winning the 2019 Young Architect Award" on their blog,

〈압구정초등학교 다목적강당〉
〈언북중학교 다목적강당〉
다목적강당의 구성은 참 간단하다. 아니,
복잡할 수가 없다. 평면으로 설명한다면 농구
경기가 가능한 코트를 가운데 두고 한쪽에는
출입구를 포함한 코어를, 그 반대편에는
강단을 둔 다음 발주처에서 요구한 규모를
대입하면 설계의 90퍼센트는 결정된다.
수학적으로는 그렇다는 말이다. 겉에서
보면 덩치가 커서 무언가 있을 성싶어도
막상 뚜껑을 열면 속이 텅 빈 들통처럼
너무 뻔해서 뜯어보고 자시고 할 것도
없다. 공사비는 또 얼마나 박한가. 오죽하면
천장의 전등 배치가 규칙적이지 않은
이유가 금액에 맞추어 정해진 개수대로만
전등을 설치하려고 했기 때문이라고 했다.
무언가 어색하다 싶으면 어김없이 "공사비
때문에"라는 대답이 돌아왔다. 나중엔

질문자가 직접 답해도 될 지경이었다. "아마
공사비 때문일 거야."라고.

이처럼 젊은 건축가의 재능을
표현할 만한 잉여 혹은 여지가 없다시피 한
상황에서 아이디알의 선택은 남달랐다.
특별함을 만들기 위해 속이 빈 들통을
찌그러뜨리거나 어딘가를 삐쪽하게 만들거나
자빠뜨리는 방식을 선택한 것이 아니라
들통은 들통다워야 하므로 속이 텅 비어야
옳다고 생각한 것이다. 그 들통으로 만두를
찌기도 하고 빨래를 삶기도 하고 사나흘 간
곰탕을 끓이는 등 쓰임이 바뀌면 약간의
설거지만으로도 언제든 새롭게 쓰이는
숙명을 수긍한 것이 바로 건축가의 의도이며
설계의 실마리였다. 그래서 과하지 않았고
또한 의젓하다.

〈매곡도서관〉
완만한 경사를 이루고 있는 대지는 좁고 긴
형태로 인근 하천의 흐름을 따른다. 마을은

a year after writing "Participating
in the 2018 Korean Young Architect
Award". 臥薪嘗膽[4], 捲土重來.[5]

YOUNG ARCHITECT - THEIR HOUSE

<APGUJONG ELEMENTARY SCHOOL
MULTIPURPOSE AUDITORIUM>
<EONBUK MIDDLE SCHOOL
MULTIPURPOSE AUDITORIUM>
The composition of a multipurpose
auditorium is very simple. It's
impossible to complicate. Described
as a floorplan, you put a court for
a basketball game at the center,
with the exit on one side, and the
stage on the other, according to the
capacity required by the commissioning
body. This determines around 90
percent of the design, at least in
terms of numbers. Its huge mass,
seen from the outside, might seem
to have something special, but once
you open it up, it's like an empty
bucket, so obviously empty that you
don't even have to open it up. Then,
consider the stingy construction
budget. Even the irregular placement
of the lighting on the ceiling was
apparently due to their attempt to

find a way to fit everything within
the budget. Whenever I felt like
something looked a little off and
asked them, the response was always
"construction costs." Towards the
end of the tour, I felt I could just
answer my own questions with, "it's
probably the construction costs."

As such, in situations where it
seemed there was no leeway or outlet
for the young architects to express
their talent, the choices of IDR was
precise. They didn't attempt to make
the empty bucket special by squashing
it, or rendering it more pointy, or
collapsing it into the ground; they
simply came to the conclusion that
a bucket is more beautiful when it
looks like a bucket, and that it
should rightfully remain empty. The
intention of the architect, and the
point of departure for their design
was their acceptance of the destiny
of the bucket - a tool which can
always reliably be used for something
else given a little washing up: to
steam dumplings, to boil laundry,
or to brew a bone broth for several
nights and days. In short, the design
avoided excess and exudes maturity.

그 반대편 낮은 경사지에 있어 사람들이
도서관을 이용하려면 거슬러 올라가야 한다.
도서관 대지뿐 아니라 주변 일대가 경사지다.
그러므로 '경사진 대지'라고 하기보다는
'경사진 동네에 도서관이 있다'고 하는 것이
옳다. 그런 조건에서 건축가가 거듭 강조한
핵심 장치는 건물 내부에 있는 경사로다.
낮은 곳에 있는 현관을 지나며 시작되는
경사로는 도서관 한복판을 꿰뚫으면서
반 층 올라간 반대편에 도달하고 가운데에
있는 열람실을 감싸며 다시 올라가면
그다음 층으로 연결되는 방식으로 공간
전체를 이어준다. 이를 두고 '경사로를
만들었다'고 말할 수도 있겠지만 본래부터
있던 경사지를 성형하지 않고 그 위에
'건물을 얹은 결과'라고 말하는 것이 더 옳을
것이다. ‹매곡도서관› 또한 만들었다기보다
있는 그대로 수긍해서 나온 결과로 그것이
바로 건축가의 의도이자 설계의 실마리였다.
그래서 과하지 않았고 또한 의젓했다.

그제야 풍문으로 들은 젊은건축가상
심사위원들의 귓속말이 무엇을 의미하는지
알 것 같았다. 누군가 던진 그 말의 출처는
이승환의 구부정한 허리춤이나 숭숭한
머리숱을 두고 한 말이거나 그들의 나이가
젊은건축가상 응모의 한계인 45세에 꽉
차서가 아니었다. 또한 블로그에 실패기까지
쓰며 연거푸 두 번을 젊은건축상에 응모한
노익장에 대한 예우나 슬하의 자녀가
하나도 둘도 아닌 셋이라서도 아니었다.
프레젠테이션에서 표현한 그들의 건축이,
건축 속에 도사린 그들의 태도가 또래보다
한층 원숙하다는 의미였다. 빈 들통과
경사진 대지를 대하는 의젓함을 심사자들도
알아본 것이다.

180

<MAEGOK LIBRARY>
The site, which establishes a smooth
slope, follows the line of a nearby
stream in its long narrow form. The
village lies on the opposite side on
a low-lying slope, so that people
have to walk against the slope to
reach the library. It's not just the
library, the entire neighborhood
is riddled with slopes. Hence, it
would be more correct to say that
"the library exists within a hilly
village" rather than describing
the site as a "hilly site". Within
these conditions, the architects
decided to establish a slope on the
inside of the building – a decision
which they repeatedly emphasize. The
slope, which starts when passing the
low lying lobby, pierces through
the center of the library, mounting
half-a-floor up on the other side,
wrapping around the reading room at
the center, and moving upward again
to connect to the next floor, finally
tying together the entire space.
We could look at this and say that
they "made a slope," but it would
be more accurate to say rather that
they surgically altered the original

slope, by "placing the building"
on top of it. The intention of the
architect and the point of departure
for the design of ‹Maegok Library›
can also be seen as the result of
embracing whatever existed before,
rather than creating something new.
Again, in avoiding excess, the design
exudes maturity.

"THEY'RE NOT YOUNG."
It is now possible to figure out
the comments whispered by the Young
Architect Award adjudicators, that
we have come to learn of through the
grapevine. The cause of this comment
would not have been because of the
slightly bent back and sparse hair
of Seunghwan Lee, or their common
age of forty-five, at the verge of
exceeding the eligible age limit. It
also wouldn't have been due to their
vigor despite their years, or about
applying twice to the Young Architect
Award consecutively, all the while
going to the point of writing about
their failure on their blog, or the
fact that they have not one, nor two,
but three children to take care of.
This phrase was intended to portray

젊은 건축─그들의 절망

"아니 이럴 수가……"

〈매곡도서관〉에 들렀을 때 알루미늄 루버에
입힌 필름이 검붉게 변색한 것을 보고
아이디알의 이승환은 이내 깊은 슬픔에
빠졌다. 루버에 고개를 기댄 채 표면을
만지던 그는 통곡의 벽에 선 바울처럼
침통한 모습이었다. 이는 다른 건물에서도
계속되었다. 시공자의 설계 도면 오독으로
기둥이 벽을 이탈한 모습에서, 허리 벽
상단에 나타난 잘려진 기둥의 흔적을
매만지며, 안전을 지나치게 고려하다 보니
난간이 아니라 철창이 되어버린 연결
복도에서, 다목적 체육관의 수직 창문이
검은색 필름지로 덮인 것을 보고, 걸레받이가
설계 도면과 달리 처리된 것을 보고,
현관 홀에 존재했던 대기 공간이 사라져
열람실 안쪽이 훤히 노출된 것을 보고,
아트리움 중앙에 있는 수직 계단의

핸드레일 아래 숨긴 T5등이 설치된
각도 때문인지 계단을 내려가는 사람을
눈부시게 하는 것을 보고, 램프의 끝에서
이어진 부출입구가 자물쇠로 잠긴 것을
보고, 체육관으로 유입되는 공조 덕트가
대책 없이 노출된 것을 보고, 투명해야
할, 그래서 관람석처럼 보여야 할 창문이
폴리카보네이트polycarbonate로 대체된 것을
보면서 말이다. 그야말로 '절망의 순례'였다.

그런 절망의 배경에는 야박한
공사비, 감리를 할 수 없었던 건축가의 지위,
몰지각한 공무원의 의식, 건축가의 전문성을
인정하지 않는 풍토가 있는 것도 사실이다.
그럼에도 그들의 절망에 온전히 공감할 수
없는 이유는, 그것이 최종 결론이라면
'희망은 어디서 찾아야 하나?' 하는 되물음과
이에 대한 내 경험이 겹쳐서다. 허구한 날
멱살을 잡히는 집수리 세계에 비하면 새
발의 피라는 것이 아니라 "공사비가 충분히
있다는 건축주를 만날 날이 있을까?"

how the architecture that they had
described in their presentation,
the approach distilled within their
architecture, was a degree more
mature than that of their peers. The
adjudicators recognized a maturity
in the way in which they dealt with
empty buckets and sloping sites.

YOUNG ARCHITECTURE — THEIR DESPAIR

"I CAN'T BELIEVE IT……."
Seunghwan Lee fell into a deep
sadness on a visit to the ‹Maegok
Library› as he took in the sight
of the dark brown discolored film
applied to the aluminum louver. With
his head leaning against the louver,
his hands running over its surface,
he looked as full of heartache as
Saint Paul standing at the Wailing
Wall. This continued in all of the
projects: Where a column had been
misplaced away from the wall due to
a contractor misreading the drawing;
As he was touching the remnants of
a column that had emerged, cut off,
at the upper part of the waist-high
wall; In the connecting corridor
installed with crossbars instead of

a balustrade due to an excessive
consideration for safety; The sight of
the multipurpose auditorium vertical
windows wrapped in black film; where
the baseboard had been installed
differently from the drawings; Where
the waiting corner in the lobby hall
had disappeared and its interior
widely exposed; Where the T5 lamp
hidden under the handrails on the
vertical stairs at the center of the
atrium was installed at an angle
which makes people descending the
stairs shield their eyes from the
light; Where the secondary exit
continuing at the end of the ramp had
been locked with a padlock; Where the
ductwork introduced in the auditorium
had been exposed without any resolve:
Where the window, which should have
been transparent, to mimic the stands
of an auditorium, had been replaced
with polycarbonate. It was, in all
honesty "a pilgrimage of tragic
despair".

Lying behind this despair is
the stingy construction budget,
the architect's lack of authority
to supervise the project, the
ignorance of the public officials,

"지각 있는 공무원을 만났더라면 좀 더 나아졌겠지만 몰지각한 공무원은 어딜 가든 또 만날 텐데 그때마다 성토의 글을 페이스북에 써야 하나?" "건축가의 전문성을 인정받지 못하는 것이 속상하지만 한편으로 책임은 덜어지지 않나?" 같은 되물음 말이다. 그런 절망의 요인들은 특별하다기보다 오히려 일상이 아니냐는 것이다. 더군다나 당장 해결할 수도 없는 문제다. 교육청 공무원이 집단으로 경질되고 GDP가 수년 안에 세계 1위가 되거나 건축가의 전문성에 대한 대우가 영국왕립건축가협회Royal Institute of British Architects, RIBA와 동급이 되기에는 양측 모두 미달이어서다. 그리고 어쩌면 그들의 절망은 오해일 수도 있다. 가령 다목적강당에서 수직 창으로 들어오는 강렬한 서향 빛으로 인한 눈부심이 활동을 방해하는 상황에서 해결 방법을 찾을 능력이 부족하고 그렇다고 서쪽에 면한 건물 배치를 바꿀 수도 없기 때문에 창을 막은 것이지

건축가의 설계 의도를 무시해서가 아니라는 것이다. 몰지각해서가 아니라 불가피했거나 건축가와 상의할 생각 자체를 못 했기 때문일 수 있다. 반면 건축가는 천창에서부터 이어지는 빛의 흐름을 바닥까지 연결해 건축적 개념을 증명하고 그것을 외피와 구분하는 데까지 확장하고 싶었을 것이다. 그럼 서쪽에 면한 수직창의 부작용을 몰랐을까? 만약 몰랐다면 감각이 예민하지 못한 것이고, 알고도 했다면 개념이라는 추상적 덫에 구속되었거나 외면한 것이리라. 절망의 원인이 흉측한 필름지를 붙인 사람이 아닐 수 있다는 말이다.

공사비도 마찬가지다. 현상 설계 참여자들이 흔히 하는 이야기가 설계해놓고 견적을 뽑았더니 예산을 초과하는 바람에 재료를 바꾸느라 애를 먹었다는 것이다. 〈매곡도서관〉도 그랬다는데 직접 보니 노출된 구조체만으로도 그 원인이 짐작되었다. 구조 계산의 경험을 대입해보면 경사지를

and social norms and traditions that do not recognize the expertise of the architect. Even so, one cannot entirely empathize with their despair, because, if that were the final conclusion, I would have had to ask myself "where do we find hope?" while scenes of my own experience overlapped. I don't mean to say that what they're saying sounded trivial, but compared to the world of restoring houses, where I'm grabbed by the collar practically every other day, it made me question, time and again, "will they ever meet a client with a sufficient construction budget?" "it would have been wonderful if they had met a highly aware public official, but given the fact that there are ignorant public officials everywhere, are they going to denounce such practices every time on facebook?" "while it's a pity that there is no recognition for architect's expertise, doesn't this mean less responsibility?" These questions point to how these elements of despair are not at all unique, but a significant part of our daily lives. Moreover, these are issues

which can't be solved straight away. It is out of the abilities of both parties, as there will be no mass reshuffling of the Office of Education officials unless the national GDP surges to first in the world, or that the treatment of architects suddenly becomes similar to that of the Royal Institute of British Architects. And perhaps their despair might also be a misunderstanding. Perhaps the people at school couldn't figure out how to resolve the glare of the sunlight which enters through the vertical window of the multipurpose auditorium disrupting activities, and since they couldn't change the orientation of the building towards the West, they had blocked off the window because of this, rather than with the intention to ignore the architect's direction. It wasn't because they were ignorant, but rather because it was inevitable and because they hadn't even thought of consulting the architect about this. On the other hand, the architect would have wanted to continue the flow of light entering in from the skylights all the way to the floor, to prove their architectural concept,

염두에 둔 구조 프레임인가 싶었고 공간 성형을 위해 불필요한 부분까지 콘크리트로 처리되었음을 확인했다. 가뜩이나 부족한 공사비에서 구조체의 성형에 지나치게 많은 비용이 투입되면 정작 필요한 부분이 생략되거나 미흡하게 처리될 수밖에 없다. 어쩌면 절망의 대상은 부족한 공사비가 아닐 수 있다는 것이다.

### 젊은 건축가—그들의 지금

보아하니 그들은 공공 건축을 계속할 것 같지 않은 눈치다. 교육청 블랙리스트에 올라가 있어서 하고 싶어도 못하는 데다가 경쟁률도 높아졌다고 농담처럼 말하지만 그보다는 지난 몇 년 동안 진저리가 난 것이다. 하긴 한꺼번에 당선된 〈압구정초등학교 다목적강당〉 〈언북중학교 다목적강당〉을 맡아 일하는 동안 얼마나 자주 하늘을 올려다보았을까. 얼마나 자주 심호흡을

했을까. 얼마나 많이 손아귀를 움켜쥐었을까.

"어린 아기를 데리고 하루 왕복 7시간을 길에 뿌려가며 울산에 회의하러 다녔던 〈매곡도서관〉 프로젝트도 결코 쉽지는 않았지만, 다목적강당 프로젝트에 비하면 그야말로 꽃길이었다는 생각마저 들 정도로 교육청 프로젝트들은 어렵고도 힘들었다."[4]

마음이 짠하다. 그렇지만 모조리 지나간 일로 접어두기에는 너무 아깝지 않은가. 세 번의 감리자 아닌 감리 경험, 수없는 좌절, 휘말린 필화 사건, 공사비 초과로 머리를 쥐어뜯던 나날, 어린 준희를 들쳐 업고 머나먼 울산을 왕래하던 기억, 서울 사는 건축가가 왜 지방까지 와서 우리의 일감을 빼앗느냐는 눈총, 농구공 한방이면 뽀개질 에어컨 박스의 쓸쓸함을 둘만의 기억으로 남겨둠으로써

and to have extended this as a method to differentiate the interior with the outer skin. Would they have had no idea about the side effects of the vertical window facing the west? If they didn't, it means that they were not sensitive enough. If they knew and still intentionally implemented it, they must have been trapped in the abstract due to their concept. What I'm trying to say is that their despair might not actually be because of the person who stuck that ugly sheet there.

The question of construction costs can also be applied to this. People who take part in design competitions commonly talk of how they had a hard time changing the materials after having completed the design and drew up the estimate, because it exceeded the budget. They say 〈Maegok Library〉 was also like that, and having seen the project myself, its exposed structure itself helped me imagine why. Applying my experience when seeing the structure, I imagined that the structural frame had been made to account for the slope, and I could confirm that even

the non-necessary parts had been completed in concrete to mold the space. With too much invested in the structural molding, and a budget falling short, it is inevitable that what other necessary elements had to be dropped or left incomplete. Perhaps, the reason for their despair is not the lack of a construction budget.

YOUNG ARCHITECTS — THEIR PRESENT

It doesn't look like these architects will continue to work in public architecture. While they joke that they can't even if they want to because they have been blacklisted by the Office of Education, and that competition has since risen, if they don't, it would probably be out of exhaustion from the past couple of years. It's not surprising, one can imagine how many times they would have looked up at the sky, breathed deeply, during the execution of the simultaneously nominated projects at 〈Apgujong Elementary School Multipurpose Auditorium〉 and 〈Eonbuk Middle School Multipurpose

제2의 이승환, 전보림이 또 다시 실패의 기록을 쓰게 될 것을 걱정하는 게 아니다. "좋은 경험 했다."라는 말을 남기고 두 손을 털며 이 바닥을 뜸으로써 누더기가 될 공공 건축의 미래를, 그리고 누더기가 된 공공 건축물을 이용하게 될 누군가의 대물림과 그로 인한 결과를 걱정하는 것도 아니다.

이 말을 기억할 것이다. "이제 다목적강당의 디테일을 알 것 같다."라는. 물론 그대들이 말하는 '알 것 같다'는 최초의 가정이거나 그러리라 생각하고 그려진 도면의 결과를 눈으로 확인한 뒤 교열되고 교정된 것이다. 그것이 모인 350개의 기록이 노하우처럼 저장되었을지라도 35만 개의 실패를 또 다시 앞두고 있다면 그것 또한 별거 아니다. 다시는 생각하고 싶지 않겠지만 그대들이 비토한 공무원, 그러니까 전보림의 팔뚝을 한층 굵게 만들고 푸른 콧김을 뿜게 한 그 무지막지했던 공무원, 건축가는 무시하고 학부모가 요구하는대로 색을 바꿔

칠하겠다던 교장 선생님(내가 볼 땐 그 사람들이 더 고생이 많았다. 정통 건축가를 만났으니)을 통해 이처럼 멋진 성과를 이룬 경험이 너무 아깝다는 것이다. 게다가 요즘 나오는 다목적강당 현상 설계 결과가 기실 그대들의 개념을 넘어서지 못한다고 하지 않는가? 따라서 "이제 다목적강당의 디테일을 알 것 같다."라는 말 뒤에 "그래서 이제 좀 할만하다."를 하나 더 붙인 뒤 새로 배역을 맡을 주무관과 교장 선생님을 맞으면 어떨까. 이곳에 뼈를 묻으라는 것이 아니라 새로 개척될 영토로 향하는 토대로 삼자는 것이다. 우리의 여행길에서 "그 사람은 주택 설계비를 2억이나 받았다며?" 하면서 부러워 마지않던 그 세상을 바라보며.

Auditorium›. I can imagine just how many times they would have tightly clenched their fists.

> "The project with the Office of Education was so hard that the Maegok Library Project was really smooth-sailing compared to the multipurpose auditorium projects, even though I had to attend meetings in Ulsan by travelling seven hours return to Ulsan accompanied by an infant."6

I sympathize with them. Yet……. Yet, wouldn't it be a waste to file this all away as something which has all passed? I am not worrying that all these memories - the three experiences of supervision withour being a supervisor, the endless despair, being embroiled in the internet post incident, the days of pulling their hair out about the budget, the days of binding Junhee to her back and travelling back and forth to Ulsan far away, the glare of why an architect living in Seoul would come all the way to the countryside to take away their work, the solitude of an air conditioner box that would crumble under a single basketball - will remain just as that of the two architects, and that Seunghwan Lee and Borim Jun will have to write up another record of failure. I am also not worried about them shaking their hands off the affair and leaving this field, with the words, "we had a good experience" and that this will lead to the generational inheritance of the shambles that public architecture has become, its future, and the people who are to use it.

This phrase will be remembered. "I think I understand what sort of details a multipurpose auditoriums should have". Of course, the phrase "I think I understand" is based on the results of the drawings, drawn based on preliminary hypothesis or assumptions, and then checked and proofread and edited. If one were to say that the 350 records have been saved as know-how, another 350 thousand failures is perhaps, not so great an ordeal. What I'm trying to say is that the experience of having accomplished so much, although the architects themselves may never want to think of this again, despite the public official that they denounced, that uncouth civil servant who made Borim Jun's arms a little more muscled, and made her vent with blue rage, the headteacher who, tried to repaint a wall with a color that a parent of a student insisted, not without consulting the architect. (In truth, perhaps they were the ones who had the harder times in this situation, for having met such authentic architects.) Moreover, don't they say that the design competition results for multipurpose auditoriums produced nowadays don't really go beyond the concepts produced by these architects? So, to the words "I think I understand what sort of details a multipurpose auditoriums should have." why not add another phrase "so, now we can kind of do it" to embrace the officials and head teachers in their roles to come. This doesn't mean that these two architects should bury their bones in the genre, but that this experience should form an approach towards uncharted territories. And while on that trip, we will look back to the time and the world where we once felt envious of "that person who got paid 200 thousand dollars for designing a house."

# 건축공방

# Archi Workshop

심희준은 스위스 취리히연방공과대학교에서 교환 학생으로
공부한 뒤 독일 슈투트가르트대학교를 졸업했다. 이후
렌초피아노건축사무소, 헤르조그앤드드뫼롱건축사무소,
라쉬앤드브라다취건축사무소에서 실무 경험를 쌓았고,
렌초 피아노가 설계한 광화문 KT본사 사옥의 디자인 감리
컨설팅을 맡았다. 현재 서울시립대학교 건축학과 겸임 교수와
새건축사협의회 정책 위원으로 활동하고 있다.
박수정은 광운대학교 건축공학과를 졸업하고 네덜란드
델프트공과대학교에서 에라스무스 교환 학생으로
공부한 뒤 독일 슈투트가르트대학교를 졸업했다. 이후
독일의 베니쉬건축사무소, 메카누건축사무소와 한국의
오이코스코리아에서 실무 경험을 쌓았다. 현재 새건축사협의회
정책 위원, 서울시 공공건축가로 위촉되어 활동하고 있다.

Heejun Sim is a founder of ArchiWorkshop.
He participated an exchange student program at
ETH Zürich and graduated with a Vordiploma and
Diploma at the University of Stuttgart in Germany.
He has worked in broad range of European offices,
such as RPBW (Renzo Piano Building Workshop) in
Paris, Herzog & de Meuron in Basel and Rasch &
Bradatsch (SL-Rasch) in Stuttgart. He was a design
consultant for the KT Headquarters in Seoul by
Renzo Piano. Currently, he is an adjunct professor
at the university of Seoul and a committee member
of the Korea Architects Institute.
Sujeong Park is a founder of ArchiWorkshop. She
graduated from Kwangwon University with a degree
in architecture. She participated in an Erasmus
exchange student program at Delft University of
Technology in the Netherlands, and graduated
from the University of Stuttgart in Germany
(Diploma). Since then, she has worked in broad
range of European offices, such as Behnisch
Architekten in Stuttgart, Mecanoo in Delft, as
well as Oikos in both the Korea and Wageningen
offices. She is a committee member of the Korea
Architects Institute and has also been commissioned
as a Seoul Metropolitan Government public
architect.

# 일상

# The everyday

Essay One

#사용자의삶 #불편함 #행복함 #화이트큐브망우 #기능성

ㄱ 건물 틈새에 핀 들꽃도 일상에
위안을 준다.

건축공방이 생각하는 일상, 그 일상이라는 주제는
어디에서 왔을까? 건축가인 우리에게 일상의 건축이
중요한 주제가 된 이유는 일상적으로 살아가는 공간이 곧
건축의 문제이고, 행복에 관한 것이기 때문이다. 집에서,
사무실에서, 작업장에서, 식당에서, 학교에서, 오고 가는
길에서, 하루를 시작하고 마무리하며 오늘을 어떻게 보냈고
얼마나 행복했는지 돌아본다면 어떤 답이 떠오를까? 우리의
관심사는 공간이 행복이라는 문제에 얼마나 관여하고
있는지다.

　사람들은 대부분 스트레스가 생겼을 때 그 원인을
'공간'에서 찾지 않는다. 하지만 우리가 느끼는 불편함은
공간의 환경에서 생겨나는 경우가 많다. 쓰레기봉투가
널브러져 있고, 전기선과 통신선이 뒤엉켜 있기 일쑤이다.
불안한 기분이 드는 밤길도 빼놓을 수 없다. 이와 마찬가지다.
보도와 차도가 구분되지 않는 길은 보행자와 운전자를 늘
긴장하게 하고 주차 공간이 마땅치 않아 이중 주차라도
한 날은 차 빼달라는 전화가 언제 올지 몰라 휴대폰에 신경을
쓴다. 물론 이런 일 외에도 문제라고 느껴지지 않는 문제가

#TheEverydayOfTheUser #inconvenience #happiness
#whitecubeMangwoo #functionality

The everyday as imagined by ArchiWorkshop: Why is
this theme important to us? As architects, everyday
architecture became important to us because
ultimately, the spaces we live from day to day are
an architectural concern, and also one that impacts
our happiness. Were we to look back on our day
from start to finish, on what we did and how happy
we were, at home, in the office, at the workshop
or the restaurant, at school and on the road, how
might we frame our thoughts? We are interested in
understanding the impact of spaces on happiness.
　For most people, 'space' is not the first
thing that comes to mind as a cause of stress.
However, the spatial environment can often be a
contributing factor to any discomfort we may feel.
For instance, let's imagine an alleyway strewn with
trash bags and tangled up electric and telephone
wires. One is certain to feel intimidated. In the
same way, pedestrians and drivers will always
feel nervous along streets where the sidewalks
and car lanes are not clearly demarcated, while
drivers who, failing to find parking, have parked
in front of another car must anxiously watch their
cellphones. Of course, many more instances exist
where we don't even realize that there might be
an issue. Regrettably, after a certain point in
time, these problems become a part of our everyday.

ㄱ Wild flowers blooming in
the crevices of buildings
provide daily comfort.

더 많다. 그런 일은 안타깝게도 언젠가부터인가 일상의
영역으로 들어왔다고 할 수 있다. 건축공방의 작업은
그 지점을 지나치지 않고 찾는 것에서 시작한다.

　　서울 중랑구에 지은 다가구 주택 〈화이트큐브 망우White
Cuble Mangwoo〉를 진행할 때였다. 설계를 맡긴 의뢰인과 첫
인터뷰를 했다.

　　"어떤 집을 원하시나요?"

　　"따뜻한 집이요."

　　"아! 네. 그럼요, 따뜻한 건 기본이지요.
　　그리고 또 원하는 것이 있으신가요?"

　　"무조건 따뜻한 집이요!"

그가 40년 동안 지냈다는 집을 살펴보니 그 이유를 알 수
있었다. 단열 처리가 거의 되어 있지 않은 1970년대 여느
집들처럼 단열재 없는 외벽과 바람이 숭숭 들어오는 창호로
버티는 환경에서 살고 있던 것이다. 여름에는 덥고 겨울에는
아무리 난방을 해도 집안에 온기가 남지 않는다. 그는
겨울을 나기 위해 집에서 패딩을 입고 두꺼운 양말을 신고
연탄난로까지 들였다고 했다. 매년 겨울이 찾아올 때마다
익숙한 듯 그리 살았던 것 같다. 건물이 제대로 기능하지
않아 생긴 불편함이 40년 동안의 일상을 만들었다.

The work of ArchiWorkshop starts with diagnosing
exactly when this occurs.
　　It happened when we were working on the
project ‹White Cube Mangwoo›, the multiple housing
project in Jungnang-gu, Seoul. It was during
our first interview with the client who had
commissioned the project.
　　"What kind of home would you like?"
　　"A warm house."
　　"Ah! Yes, of course, that's a given. Is there
　　anything else you would like?"
　　"Whatever it is, it must be a warm house!"
Upon observing the house he had been living in
for the past four decades, we finally understood
why. The client had been living in an environment
with little insulation, or rather, he had hung on
despite uninsulated walls and cold air piercing its
way through the windows and doors. The house was
hot in the summer and, in the winter, regardless
of how high the heat was set, refused to retain
heat. To survive the winters, the client would
wear padded coats, thick socks and even installed
a supplementary charcoal fire. He had continued
each year, at the turn of winter, as if this
were normal. This discomfort of a malfunctioning
building had created his 'every-day' for four
decades.
　　It is also problematic when architecture
intentionally imposes discomfort upon its users.
This can also be an issue when designing space. The

건축이 사용자에게 불편을 의도하는 것은 옳지 않다.
이 문제는 공간을 만들 때도 적용된다. '비가 새야 건축가가
지은 집'이라는 우스갯소리(이 이야기를 처음 들었을
때, 우리는 전혀 우습지 않았다)는 더 이상 전문가의
영역에서 존재할 수 없다. 건축 설계를 하면서 그 디자인을
뒷받침할 상세도도 제대로 그리지 않고 시공자에게 모든 걸
떠맡긴다면 사용자의 삶은 고려되지 않은 것으로밖에 볼 수
없다. 이는 전혀 전문가가 아니다.

"신은 디테일에 있다. God is in the detail."는 미스
반데어로에의 말처럼 사소하지만 섬세하게 접근하는 법은
조용하게 일상의 변화를 만들어낸다. 옆 건물에 팔만 뻗으면
손 끝에 닿을 정도로 가까워서 몇십 년 동안 창문을 제대로
열지 못했던 의뢰인의 고민을 듣고 1평 남짓한 외부 공간을
활용해 마음 놓고 열 수 있는 창을 만든다. 내외부에 이중
단열을 하고 시스템 창호를 제대로 쓰되, 늘어난 공사 비용을
고려해 외벽 재료와 공법은 단순한 것을 택하고 관리가
용이하도록 상세도를 제안한다. 빗물 때문에 외벽이 오염되지
않도록 디자인하고 시공사에도 도면과 글로 여러 번 설명한
뒤 만나서 또 당부한다.

2013년 유럽에서 한국으로 돌아왔을 때 누군가 우리가

inside joke (which we did not find at all funny
the first time we heard it) that "you can't call
it a house built by an architect if it doesn't
leak a little when it rains" should no longer
exist in this field. It goes without saying that
the architect does not care about his/her client,
if he/she delegates everything to the contractors,
without even bothering to draw detailed drawings
to support his/her design. It is impossible to
consider this person as an expert.
      Ludwig Mies van der Rohe once said, "God is
in the details." Similarly, intricate, even minute,
detailed approaches can quietly elicit change in
the everyday. After learning how the arm's length
proximity of the neighboring building had prevented
the client from opening his window properly for
decades, we created a window which could be easily
opened, using an outdoor space of no more than
a 1 pyeong (3.3m²). We proposed double internal
and external insulation and a tilt and slide
system for windows, while lessening the burden
for construction costs by using simple materials
and construction methods for the outside wall.
We also created a detailed drawing to help with
construction. The design ensured that the outer
wall would not be polluted by rain, and through
countless meetings we explained the drawings and
text several times to the contractors.
      When we returned to Korea from Europe in
2013, somebody, who heard we had studied in

화이트큐브 망우 (2014)
서울 중랑구에 위치한 다가구 주택으로
일반적인 주거 지역에 속해 있다.

WHITE CUBE MANGWOO (2014)
It is a multi-family house located in
Jungnang-gu, Seoul, and belongs to the
general residential area.

화이트큐브 망우 (2014)
발코니에서 햇빛이 쏟아져 들어와
실내를 밝게 만든다.

WHITE CUBE MANGWOO (2014)
Sunlight pours in from the balcony,
making the inside bright.

독일에서 공부했다는 소식을 듣고 '패시브하우스Passive House'에 대한 책을 써보라고 권했다. 당시에 우리는 다가구 주택을 설계하고 있었는데 옳지 않은 부탁을 받은 것처럼 마음이 불편했다. 단열 구조를 제대로 갖추지 못한 집이 절반 이상인 한국에서 패시브하우스에 관련된 책을 쓴다는 사실에 괴리감을 느낀 것이다.

　건축이란 이론과 실재의 양면을 모두 가지고 있다. 따라서 우리는 역사와 이론, 예술과 문화에 관해 늘 고민한다. 우리가 프로젝트를 진행할 때 적어도 하나 이상의 사회적 이슈를 고민해 담아내려 하는 것도 그런 이유다. 오랫동안 재개발 구역으로 지정되어 있던 곳이 지역 주민의 자발적 반대로 해제되거나 소위 집 장사의 건물로 빼곡한 동네의 배경을 건축물 하나로 살필 수 있다. 집 하나를 짓는 것으로 주변 건축과 도시의 맥락에 관한 이야기가 전달될 수 있다.

　그래서 우리는 일상을 먼저 가다듬는 일에 주목해 보이지 않는 불편함을 끄집어내려고 한다. 국내 주거 유형의 60퍼센트 이상을 차지하는 아파트를 한번 보자. 그 평면 구성은 오랫동안 발전이 없다. 불편함도 그대로 이어져 오고 있다. 대다수가 베란다를 창고로 여기고 짐을 쌓아두기

Germany, recommended that we write a book about the Passive House. At the time, we felt uncomfortable as if we had received a wrong request, as we were in the midst of designing a multiple-housing home. We felt a sense of distance at the thought of writing a book about passive housing in Korea, where more than half of the houses do not have proper adequate insulation.

　　As architects, we have always strived to consider history and theory, art and culture. Architecture marries theory with practice, necessitating that we attempt to embrace and question at least one or more social issue during a project. The construction of a single house can communicate the underlying stories of the surrounding architectural and urban context. For example, how a district was liberated from redevelopment due to the voluntary resistance of locals, or how a village came to be packed with what we call "house-seller" buildings. A single piece of architecture can tell the story of a place.

　　In this way, we focus our attention on dealing first and foremost with the everyday. We aim to tease out unseen discomforts. Let's consider apartments, which they say is more than 60% of the residential typology of Korea. The composition of the typical apartment floorplan has remained unchanged for a long time, and therefore associated discomforts have continued. Since most verandahs

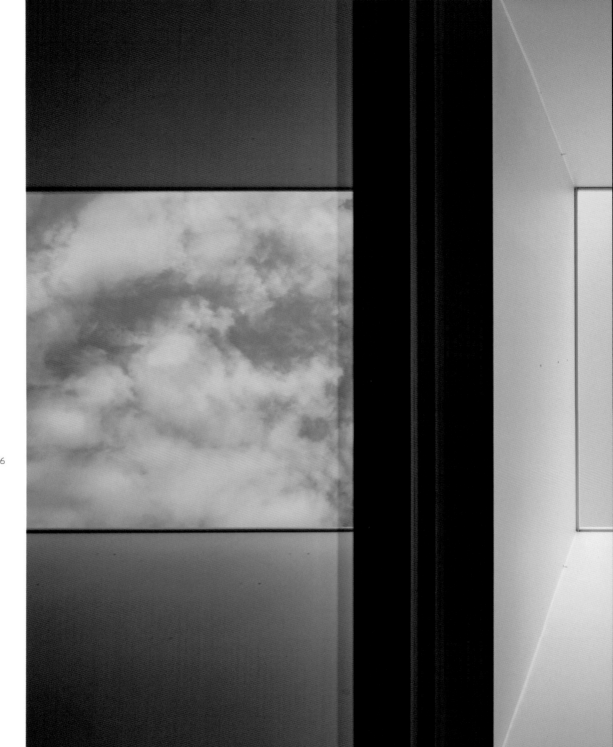

베지가든 (2014)
시원하게 열린 천창을 통해 시시각각 변하는
하늘의 표정을 느낄 수 있다.

VEGEGARDEN (2014)
Through the open skylight, users can feel
the changing look of the sky.

때문에 거실이나 방에서 바깥을 시원하게 볼 수 없다. 지하
주차장에 주차하고 차에서 내린 다음에는 어두컴컴한
통로를 지나 엘리베이터 전실까지 걸어야 한다. 이외에도
도시 곳곳에는 여전히 익숙한 불편함이 존재한다. 학교는
교육과정의 변화를 따라오지 못한 채 아직도 판에 박힌 듯
일률적이고, 화장실은 여전히 공포스럽다. (화장실이 무서워
온종일 볼일을 참는 학생도 더러 있다고 한다.) 학원 창문은
창이 없거나, 밖이 보이지 않게 짙은 코팅지로 덮여 있다.

그러나 공간의 힘은 분명히 존재한다. 미국의 정신
건강 전문가 에스더 M. 스턴버그Esther M. Sternberg가 쓴 책
『공간이 마음을 살린다 Healing Spaces: The Science of Space and Well-
Being』에는 일조량이 많아 치유가 빠른 병실, 마음을 위로하는
정원, 스트레스를 덜어주는 사무실, 영감이 솟는 연구소,
건강한 도시, 위안받는 집에 대한 이야기가 나온다. 이처럼
일상은 모든 건축 공간과 연관된다. 건축의 변화가 일상의
변화를 주도하는 것이다. 그런 이유로 우리는 다양한 건축적
프로그램과 주제에 관심이 있다. 주거, 사무실, 공공 공간,
연구 공간, 상업 공간, 글램핑Glampling, 파빌리온Pavillion,
리조트, 설치 예술 등 우리의 작업 범위는 계속 확장된다.
일상에서 더 높은 수준의 환경을 경험할 수 있도록 삶의 질을

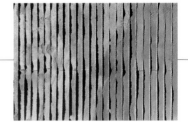

노출 콘크리트 입면에는 시간의
흐름이 포착되어 새겨진다.

are used as storage, one cannot see outside with
an open view from the living room or the bedroom.
After parking underground, one must walk through
a darkly lit pathway to the elevator. Moreover,
children's schools are all so obviously the same
without following the trend, and the bathrooms
look like scenes from a horror movie (Apparently,
some students try avoid  the bathroom all day long
because of this fear). After-school academies often
do not have windows, or their windows are covered
with dark plastic sheets blocking out the view.

However, space is clearly powerful. The book
『Healing Spaces: The Science of Sapce and Well-Being』
written by the American mental health expert Esther
M. Sternberg features stories of hospital wards
with higher recovery rates due to greater amounts
of sunlight, gardens to comfort the soul, offices
reducing stress, inspiring research labs, healthy
cities, and comforting homes. Everyday experiences
are impacted by all architectural spaces. Change
in space make changes in everyday life. Hence, we
must take an interest in a variety of architectural
programs and themes. The scope of our projects
continues to expand from housing, offices, public
spaces, research, commercial spaces, glamping,
pavilions, resorts, installation art, etc. We wish
to continue with projects which enhance quality of
life so that one can experience better environments
in the commonplace and the everyday. We believe
in the empowerment that can come from a space,

↗ The exposed concrete facade
captures the flow of time as
a shadow.

높여주는 작업을 지속하고 싶다. 우리는 공간이 주는 힘을
믿고, 그 힘은 생각보다 크다. 단순하고 강하며 서정적이고
아름답고 우리의 현재를 즐길 수 있으면서 우리의 현재가
발전하고 이어지는 지속 가능한 건축. 우리가 생각하는
일상의 건축은 그런 것이다.

and that it can strengthen us more than we can
imagine. Sustainable architecture which is simple
yet powerful, poetic and beautiful, which allows
us take pleasure in our present-day amidst its
development and continuation: This is the kind of
everyday architecture we imagine.

# 유럽

# Europe

#실무 #독일유학 #KT광화문사옥 #귀국 #개소
#백사마을프로젝트

유럽에서 10년 이상 지내는 동안 '건축에 관한 지식'만을
배운 것은 아니었다. 우리는 둘 다 독일에서 건축을 배웠다.
심희준은 호주에서 대학을 다니던 중에 독일어를 배우기
위해 머물렀던 도시 프라이부르크Freiburg에 반해 독일 유학을
선택했다. 박수정은 한국에서 건축을 전공하고 첫 번째
사무실인 오이코스코리아Oikos Korea에서 실무를 경험하면서
접하게 된 유럽 건축에 매료되어 독일로 유학을 하러 가게
되었다. 심희준은 가볍고 실험적인 건축에 관심이 많았고,
박수정은 생태적이고 친환경적인 건축과 도시 공간에
관심이 많았다. 돌이켜보면 둘에게 독일은 이 모든 것을
경험하기에 좋은 선택지였다. 독일 건축 대학 대다수가
공과대학교에 뿌리를 두고 있기 때문이다. 우리가 공부한
슈투트가르트대학교University of Stuttgart는 도시, 구조, 건축
분야에서 유명한 대학으로, 다양한 기관이나 워크숍 활동을
통해 폭넓은 건축을 경험할 수 있는 곳이었다. 또한 서로
다른 국가의 다양한 건축 이론과 가능성을 배우기에도
적합했다. 심희준은 스위스 취리히연방공과대학교Swiss Federal

#practice #StudyingAbroadInGermany
#KTGwanghwamunHeadquarters #ReturningToKorea
#openingArchiWorkshop #BaeksaVillageProject

Our decade spent in Europe was not simply about
amassing "knowledge about architecture." We both
studied architecture in Germany. Heejun Sim chose
to study in Germany after falling in love with
the city Freiburg where he had learnt German as a
university student. Sujeong Park graduated from her
BA in Korea and decided to study in Germany after
falling in love with the European architecture
she had seen while working at Oikos Korea, her
first job. Heejun was greatly interested in
light yet experimental architecture, Sujeong was
greatly interested in ecological and environmental
friendly architecture and urban spaces. Looking
back now, Germany was a good choice for both of
us to experience all of these elements. German
architecture programs characteristically start in
engineering schools. The University of Stuttgart
where we studied was a university renowned for
the urban design, structure and architecture,
allowing us to experience a wide range of
architecture through a variety of institutes and
workshops. It was also a good place to learn
about the architectural theory and potential of a
variety of different countries. Heejun Sim was an
exchange student at the Swiss Federal Institute
of Technology Zürich, while Sujeong Park was an

Institute of Technology Zürich에서 교환 학생 과정을, 박수정은
네덜란드 델프트공과대학교 Delft University of Technology 에서
에라스무스 Erasmus 프로그램으로 교환 학생 과정을 거쳤다.
우리가 지낸 독일, 스위스, 네덜란드는 유럽에서도 건축
사무소가 많이 있는 국가였다. 덕분에 학생 때부터 실무를
경험했고 졸업한 뒤에는 건축가로 일하면서 프로젝트의
아이디어를 제안하고 개념을 발전해가는 과정, 다른 팀과
협업하는 법, 건축가의 역할, 체계적 시스템 등을 다채롭게
경험할 수 있었다.

그러다가 심희준이 파리의 렌초피아노 빌딩 워크숍
인터내셔널 Renzo Piano Building Workshop International Office 에서 서울
종로구에 있는 ‹KT 광화문사옥› 설계에 참여하게 되었고,
이후 디자인 감리를 위해 한국에 돌아왔다. 2013년 가을
건축공방을 열면서 유럽에서의 경험을 바탕으로 한국의
사회와 문화를 담는 건축 작업을 이어가고 있다. 지금까지 한
이야기는 어쩌면 일반적인 유럽 유학에 관한 것일지 모른다.
그러나 우리는 그곳에서 더 가치 있는 경험을 했다.

독일에서 생활한 지 얼마 되지 않았을 때였다.
신호등이 없는 횡단보도를 지나려는데 아주 큰 트럭이 오고
있어 트럭이 먼저 지나가기를 기다리려고 멈춰 섰다. 그런데

심희준은 서울 종로구에 있는
‹KT 광화문사옥› 설계에 참여하면서
디자인 감리 업무를 맡았다.

203

exchange student through the Erasmus program at
the Delft University of Technology. Thanks to
this, we were both able to practice architecture
as students, and after graduating it was possible
to have a diverse range of experiences of working
as an architect, pitching ideas for projects, and
the process of developing these ideas, ways to
collaborate with other teams, and the role of the
architect and systems.

It was at this point that Heejun Sim
participated in the design of the KT Headquarters
in Jongno-gu, Seoul at the office of Renzo Piano
Building Workshop International Office in Paris,
and returned to Korea to supervise the design.
ArchiWorkshop was opened in the fall of 2013, and
building on our experiences in Europe, we have been
working on architectural projects portraying Korean
society and culture. What we have discussed so far
may look like the standard experience of studying
abroad in Europe. However, Europe taught us more
valuable lessons.

We had just recently moved to Germany, and we
were trying to cross the street without a traffic
light. A huge truck was coming, so we stopped to
wait for the truck to pass by first. Yet, the truck
stopped at a place far off from the stop-line at
the crossroads, and waited for us to cross. It was
a huge cultural shock to see the enormous truck
acknowledging the stop line as if this were the
most normal thing to do. Such small acts reveal an

↗ Heejun Sim took charge
of design supervision for
the construction of <KT
Headquarters> in Gwanghwamun,
Jongno-gu, Seoul.

KT 광화문사옥 (2012)
프랑스 파리에서 설계에 참여하고 국내에서 디자인 감리를
맡아 진행한 프로젝트.

KT HEADQUARTERS (2012)
The project in charge of the design
supervision.

횡단보도 정지선에서 한참 떨어진 곳에 트럭이 멈추더니
우리가 먼저 지나가기를 기다리는 것 아닌가. 당연하다는
듯이 정지선을 준수하며 서 있는 거대한 트럭을 보고 우리는
문화 충격을 받았다. 이런 작은 행동의 이면에서 사회를
이루는 구성원의 동의가 드러난다. 사람이 우선이고, 강한
것이 약한 것을 보호해야 한다는 사회적 합의 말이다. 그
외에도 개찰구가 따로 없는 대중교통 시스템과 취업을
준비할 때 졸업 증명서를 제출한 적이 없다는 점만 봐도
사회에서 상호 신뢰 관계가 상당히 높은 수준으로 지켜지고
있음을 알 수 있었다.

우리는 유럽 여러 나라의 사회 기반 시설을
이용하면서 자연스럽게 도시를 경험하고 법규와 정책
테두리에서 건강한 건축이 나오기 위한 합리적인 과정을
경험했다. 가설계라는 단어가 존재하지 않고 설계비에 대한
합리적 지침이 세부적으로 정해져 있다. 이런 가치는 당장은
눈에 보이지 않아도 건축 설계의 완성도를 높이는 데 중요한
역할을 한다. 중소기업이 제공하는 다양한 건축 자재 산업과
그 정보의 전달 과정은 건축가와 제조사 사이의 긍정적 협업
동반 효과를 만든다. 작업을 시작할 때 건축가와 기술자가
대화를 나누며 작업의 방향성을 함께 고민하고 건물을

inherent social consensus. It is a social consensus which speaks of putting people first, and that the strong must protect the weak. Similarly, public transport systems without any ticket inspection gates, and the fact that we never had to submit our diplomas when preparing to get a job suggested to us that mutual trust was operated at a high level.

Utilizing a variety of social infrastructure facilities in many European countries permitted us to naturally experience cities and the logical process necessary for healthy architecture to be born from a framework of regulations and policies. Instead of the term 'preliminary design,' rational guidelines actually exist to define design fees. These values may not be immediately apparent, yet are crucial in improving the level of completion for architectural design. The many architectural material industries provided by small and medium companies, and the process of communicating such information creates a positive collaborative relationships between architects and manufacturing companies. Upon starting on a project, the architect and the technological expert share conversations to think deeply about the direction of the project together, and this creates an atmosphere in which everyone involved in the making of the building build on each other's efforts to "challenge ourselves together."

Sujeong Park participated in the Marco Polo Wohnturm residential project built in Hamburg,

만드는 사람 모두 "같이 도전해보자."라며 뜻을 함께하는 분위기로 넘친다.

　　박수정은 독일의 베니쉬건축사무소Behnisch Architekten에서 일했을 때 함부르크의 마르코폴로타워Marco-Polo-Wohnturm 주거 프로젝트에 참여했다. 독일에서는 흔치 않은 15층 규모의 주거 타워였다. 완공 축하 파티가 열리던 날, 함께 모인 건축가, 기술자, 발주처 그리고 관계자는 성공적 협업을 축하하며 서로에게 감사함을 표현했다. 건축가가 시공 마지막 단계까지 참여했음은 물론이다. 참 이상적인 모습이라고 생각했다.

　　우리가 생각하는 전문가로서의 건축가는 건축을 만들어가는 중심에 있으면서, 좋은 건축을 만들기 위해 미학과 기능의 균형을 잡는 위치에 있어야 한다. 또한 디자인의 영역을 넘어 각 분야의 협업 균형을 잡는 역할도 해낸다. 공간이 가지는 힘이 크기 때문에 사회적 책임도 크다. 그런 이유로 우리는 건축가가 처음부터 끝까지 모든 과정에 참여할 수 있는 환경을 만들면서 작업을 진행한다. 건축주, 건축가, 시공사 사이의 수평적 관계 맺기와 배려심 역시 좋은 건축을 만들기 위해 꼭 필요하다. 이런 환경이 만들어져야 설계 과정 자체가 무엇보다 중요한 이슈가 되고, 결과물 또한

ↄ 〈마르코폴로타워〉, 박수정이 독일 유학 중 참여했던 프로젝트다.

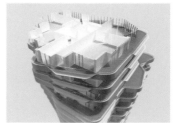

among the projects of Behnisch Architekten, the German architecture office. It was a 15 floor residential tower, a rarity in Germany. On the day celebrating its completion, the architects, technological experts, commissioners and related staff gathered together to celebrate a successful collaboration and to express thanks to each other. The architect participated until the very last stage of the construction. We thought that this was a very ideal scene.

　　When we envision the architect as an expert, when envision the architect located at the center of creating architecture, striking a balance between aesthetics and functionality. The social responsibility of the architect is great because of the enormous impact that spaces have. As such, we carry out our projects by building an environment in which the architect can participate from start to finish in all processes. Establishing equal relationships and understanding between the client, the architect and the contractor is also integral to creating good architecture. This is because we have learnt that it is only in these conditions that the design process can be prioritized, and this in turn produces good results. We wish to practice what we have learnt.

　　Our participation in the 〈Baeksa Village Project〉 in Junggye-dong, Nowon-gu, Seoul can be likened to the positive architectural process we experienced in Europe. We were nominated for

↗ Marco-Polo-Wohnturm project participated in Sujeong Park when she was in Germany.

백사마을 프로젝트
총 여섯 개 회사가 컨소시엄을 이루어 진행 중이며
2025년 완공을 목표로 두고 있다.

THE BAEKSA VILLAGE PROJECT
The design process is holding by six firms
as an consortiums and are expected to be
completed in 2025.

백사마을 프로젝트
조감도. 총 2,000세대 규모의 아파트 단지를 구릉지에
조성하는 것이 과제다.

THE BAEKSA VILLAGE PROJECT
The perspective view. The project is
establishing an apartment complex of 2,000
units.

좋을 수 있다는 것을 배웠기 때문이다. 우리는 그 배움을
실천하고자 한다.

2019년 현재 진행하고 있는 서울 노원구 중계동
〈백사마을 프로젝트〉는 유럽에서 직접 경험한 긍정적인
건축 과정과 닮았다. 국제 지명 현상 공모에 당선된
프로젝트인데 2,000세대 규모의 아파트 단지를 조성하는
것이 과제다. 설계 사무소 다섯 곳과 조경사무소 한
곳이 모여 컨소시엄(솔토지빈건축사사무소, 건축공방,
매트건축사사무소, 가아건축, 창조종합건축사사무소,
감이디자인랩)을 구성했고 머리를 맞대 프로젝트에
힘을 쓰고 있다. 건축공방은 이 프로젝트를 통해 여러
건축가와 함께 일하는 또 다른 방식을 배우고 있다.
솔토지빈건축사사무소의 대표 조남호 건축가가 이끄는
설계의 협업 과정, 각각의 팀이 선보이는 다양한 건축 언어와
접근법이 우리의 영감이 된다. 건축에 대해 의논하고 큰
주제와 상세를 함께 고민하는 과정이 서로에게 자양분이
되고 있다. 서로의 의견을 받아들이는 열린 자세와 비판하고
격려하는 전문 과정이 있어 감사하다. 이런 작업이 유럽이
아닌 한국에서 이루어지고 있어 더욱 그렇다.

↘ 〈백사마을 프로젝트〉 모형. 서울
노원구 중계동에 조성될 예정이다.

this project through an international design
competition, and tasked with establishing an
apartment complex of 2,000 units. Five architecture
offices and a landscape office gathered to form a
consortium, composed of Solto Jibin Architects,
ArchiWorkshop, MAT Architects, Ga.A Architects,
Chang-Jo Architects, Gami design lab, to put
our heads together and to work hard. Throughout
this project, ArchiWorkshop has learned ways to
work with many different architects. The design
collaboration process led by Namho Cho of Solto
Jibin Architects, the diverse architectural
language and approaches featured by each team, all
inspire us. The process of discussing architecture,
and the process of pondering large themes and
details has become a way to cultivate each other's
thoughts. We are thankful for this professional
process in which we embrace each other's opinions
with open hearts and critique and encourage each
other. This project is even more meaningful, taking
place as it does in Korea rather than in Europe.

↗ The model of <Baeksa
Village Project>. The project
site is Junggye-dong, Nowon-
gu, Seoul.

# 생존

# Survival

Essay Three

#젊은건축가강연시리즈 #모바일라이브러리프로젝트
#공공프로젝트 #하늘을담은집

2015년 건축계의 노벨상인 프리츠커상 Pritzker Architecture
Prize을 수상한 독일 건축가 프라이 오토 Frei Otto는 그 해에
89세로 생을 마감했다. 마지막 인터뷰에서 그는 "건축은
생존 Architecture is Survival"이라고 말했다. 건축가에게 생존은
어떤 의미일까? 노장 건축가의 마지막 인터뷰에 등장한 이
단어는 2014년 한국 건축계에도 등장했다. 열 팀의 젊은
건축가와 서울 소재의 다섯 대학에서 모인 4학년 건축학과
학생 200명이 한 달에 한 번 모여 진행한 '2014 한국의 젊은
건축가 강연 시리즈 10by200'에서다. 건축공방을 시작한 지
정확하게 1년이 된 시점이었고, 다른 아홉 팀 역시 대부분
사무실을 연 지 5년 미만이었다. 건축가 열 팀이 한 달에
한 번씩 차례로 강연을 진행했고 학생들은 다양한 후기를
남겼다. 손님을 초대해 이야기도 들었던 것으로 기억한다.
어떤 건축가는 좀처럼 보기 어려운 건축 현장에 관한
내용을 프레젠테이션으로 소개했고 또 다른 건축가는
화장실도 없던 열악한 주거 환경의 개선 사례를 보여주었다.
설계 내용의 좋고 나쁨을 떠나 이 강연 시리즈에서는

#YoungArchitectLectureSeries #MobileLibraryProject
#PublicProject #AHouseLiketheSky

The German architect Frei Otto, recipient of the
2015 Pritzker Architecture Prize (the Nobel Prize
of the architectural field) died the same year at
the age of 89. In his last interview, he said that
"Architecture is survival." What does survival
mean to an architect? This word also appeared
in the last interview of an elderly architect
who emerged in the Korean architectural scene in
2014. This was at the 2014 Korean Young Architect
Lecture Series 10by2000, a monthly proceeding
attended by 200 senior architecture students from
five universities in Seoul with 10 teams of Young
Architects. This was exactly a year after we had
opened ArchiWorkshop, and the other 9 teams were
also mostly firms of less than five years. Each
of the ten architect teams took turns to lecture
and the students left a variety of comments. We
also remember that guests were invited to share
their thoughts. Some architects introduced content
about the hard to come by site of an architectural
project as a presentation, and another architect
showed us case studies of improving a desolate
residential environment that didn't even have
toilet facilities.
    Whether the contents were good or bad, the
lecture series produced a critical view on young
architects (especially "young"). At the time, the

젊은 건축가에 대한(특히 '젊은'에 대한) 비판이 나왔다. 강연에서 나온 '생존'이라는 표현을 듣고 누군가는 날이 선 검처럼 혹은 뜨거운 감자처럼 느꼈을 것이다. 특히 젊은 건축가들이 생존에 급급해 건축이라는 (고상한) 문화를 만들지 못하고 있다는 데 대한 우려의 목소리가 있었다. 먹고 살기에 바쁜 '생존하는 건축가'로 '젊은 건축가'를 바라보고 있는 것이었다. 우리는 당시에 이런 일련의 상황과 소셜 네트워크 상에 떠돌아다니는 이에 대한 단상을 지켜보며 언젠가 생존에 대한 우리의 생각을 글로 써야겠다고 생각했다. 그리고 지금 그 이야기를 하려고 한다.

먼저 생존의 의미를 살펴보자. 일차적으로 생존은 살아남는 것이다. 그것도 극한 상황에서 살아남는 것인데, 남에게 피해를 주는 것이 아니라면 어떤 생존이든 환영하고 격려하는 마음이 생기기 마련이다. '내가 그 상황이었다면 어땠을까.' 하는 측은지심惻隱之心이라 할 수도 있겠다.

2013년 사무실을 열고 맞은 첫 겨울에 서울 은평구에 있는 서울혁신파크의 ‹모바일 라이브러리› 프로젝트가 들어왔다. 우리에게는 첫 번째 공공 프로젝트였다. 파빌리온Pavilion 네 개를 짓는 작업이었는데, 규모가 작았고 계약에서 설계비 조건이 좋지 않았지만, 크게 상관하지 않고

term "survival" mentioned at the lecture had struck a nerve with some. These critics were particularly concerned that young architects were not succeeding at building a refined culture of architecture due to their urgency to survive. They thought that the term "architects in survival," with an urgent need to earn one's keep, was equal to "young architects." After observing those events and the personal opinions floating around social media platforms at the time, we thought that we must write about our thoughts regarding survival in the future. And now we wish to talk about this.

First, let's look at the meaning of survival. Primarily, survival means to remain alive. This also means to remain alive in extreme conditions, and as long as it does not harm anyone else, all forms of survival should be welcomed and encouraged. This could also be seen as putting oneself in someone else's shoes, and saying "what would it have been like if I were in that situation."

The first winter we welcomed after opening the firm in 2013, a commission arrived for the <Mobile Library> in the Seoul Innovation Park, Eunpyoung-gu, Seoul. This was our first public project. It was a project in which we had to build four pavilions, and despite it being a small-scale project with less than satisfactory design fees, we decided not to pay too much attention to that, and take on the project. This was due to

일을 맡기로 했다. 일을 시작할 때 그 규모보다는 합리적
절차와 작업의 방향성 그리고 함께 만들어갈 사람들의
팀워크를 보고 결정하는 건축공방의 철학 때문이었다.
우리는 이 프로젝트를 진행하며 열악한 조건을 치열하게
고민하고 오히려 이를 부각해 건축을 완성하는 새로운
개념에 집중했다. 그러면서 우리의 삶과 맞닿은 모든
공간에서 긍정적 혹은 부정적 요소를 새롭게 발견했다.
이는 ‹모바일 라이브러리›의 파빌리온에서도 그랬고 양천구
‹신월7동주민센터› 리노베이션 설계를 할 때도 그랬다.
모든 프로젝트에서 낮은 설계비와 시공비의 벽을 넘어
대중과 맞닿는 건축이라는 플랫폼이 미약하게나마
이루어지기를 바랐다.

　　　서울 중랑구 망우동의 ‹하늘을 담은 집›에서는 사선
일조로 생겨나는 공간과 형태에 대해 고민했고, ‹21평
아파트 리노베이션›에서는 베란다에 대한 해석을 담았다.
서울 종로구의 세운상가 ‹씨드 Seed:s›에서는 외기와 직접
접할 수 없는 환경을 쾌적하게 풀어나가려고 노력했다.
이렇게 소위 말하는 열악한 조건 때문에 생겨난 건축물
자체에 생존이라고 이름을 붙이는 것이 아니라 그 과정에
생존이라는 의미가 들어가 있다.

ArchiWorkshop's philosophy that decisions to take
on a project should depend on the rational process,
the direction of the project and the creative team
rather than its scale. After reflecting long and
hard on the difficult conditions of the project,
we were able to utilize them to create a new
concept. Through all of these spaces that we came
into contact with, we were able to newly discover
positive or negative elements. This was true not
only in the pavilion of the <Mobile Library>, but
also during the renovation design of the <Shinwon-
7-dong Community Center> at Yangcheon-gu. In all of
these projects we hoped to establish, even if only
faintly, a platform of architecture in contact with
the public, by overcoming the low design fees and
construction fees.

　　In <House Embracing Sky> of Mangwoo-dong,
Jungnang-gu, Seoul, we deeply considered the space
and form established by the diagonal daylight
regulations, and the 70m² apartment renovation was
filled with interpretations about the veranda. In
the Saeun Sangga <Seed:s>, in Jongno-gu, Seoul, we
strove to create a pleasant environment which could
not be directly in contact with the open air. As
such, it is not because of the so-called difficult
conditions that we attach the name of survival to
the architecture, but rather because the concept of
survival is instilled within the process itself.

　　Korean architecture must ask and continue
to challenge itself with the question "How will

↗ The model of <Seed:s>
located at Sauna Sangga. The
architects strove to create
a pleasant environment which
could not be directly in
contact with the open air.

모바일 라이브러리-멤브레인 파빌리온 (2016)
멤브레인 천을 활용한 막 구조의 파빌리온이다.
서울혁신파크 안에 설치되었다.

MOBILE LIBRARY-MEMBRANE PAVILION (2016)
Pavilion designed with Membrane.
It was located in the Seoul Innovation Park.

모바일 라이브러리-미라지 파빌리온 (2016)
반사도가 높은 재질로 입면을 마감해 파빌리온이
도시 풍경을 입게 되었다.

MOBILE LIBRARY-MIRAGE PAVILION (2016)
The façade is surrounded by highly reflective
material, which embraces the scenery of the
city.

모바일 라이브러리-파이프 파빌리온 (2016)
가늘고 긴 파이프가 형형색색 옷을 입고 파사드를 화려하게
수놓았다. 파이프 문을 열면 새로운 공간이 나타난다.

MOBILE LIBRARY-PIPE PAVILION (2016)
The pipe covered the facade of the box. A new
space appears when you open the pipe door.

하늘을 담은 집 (2016)
사선 일조로 생겨나는 공간과 형태에 대해
고민한 프로젝트다.

HOUSE EMBRACING SKY (2016)
The architects deeply considered the space
and form established by the diagonal daylight
regulations.

건축가로서 스스로 생각의 틀을 깨며 작업하는 것. 우리만의
스타일을 만들기보다는 대지나 내용의 고유한 특성에
집중하고 완성도 있는 작업에 집중하는 것. 자기가 확신하는
디테일을 위해 일대일 목업 작업을 하는 것. 과거의 기술에
멈추지 않고 새로운 것을 받아들이는 것. 앞으로 한국
건축은 '어떻게 생존의 길을 갈 것인가?'라는 물음에 답하고
끊임없이 다시 묻는 것. 우리에게 생존은 바로 이것이다.

  우리는 하나의 프로그램이나 용도에 갇히지 않고
다양하게 도전하고 있다. 이것이 장점이기 때문에 새로운
프로젝트를 할 때마다 그에 맞는 새로운 프로그램과 개념을
창작하고 싶다. 여러 이슈를 담고 함께 고민할 수 있도록
우리에게 도전의 기회가 올 때를 놓치지 않으려고 한다.
'쇼 미 더 건축!Show Me The Architecture!'

we go about this path of survival?" Working by
breaking out of one's own framework of thought
as an architect, concentrating on the unique
characteristics of the site or the content and
focusing on completion rather than asserting one's
own style, making life size mockups to affirm the
details one believes in, accepting the new rather
than limiting oneself to the technology of the
past. For us, this is what survival means.

  Architects can challenge themselves diversely
rather than being trapped within a single program
or use. This advantage drives us to create new
programs and concepts each time we work on a
project. We try not to miss opportunities when
these challenges come upon us, so that we can
embrace a variety of issues and try to work them
out together. Show me the Architecture!

가평바위숲 '온더락' (2015)
반사도 높은 입면의 재료는 주변 풍경을 적극적으로
건물에 입힌다.

ROCK FOREST 'GLAMPING ON THE ROCK' (2015)
The glossy materials on the facade let the
landscape cover the building.

# 2019 젊은건축가상

# 2019 Korean Young Architect Award

Essay Four

#레드스퀘어하우스 #주택 #화이트큐브망우
#방배동지층사무실 #건축공방연희동사옥

2013년 당시 거주하던 24평 아파트의 작은 방에서 건축공방
사무실을 열었다. 그렇게 한국에서 건축을 시작했다.
지금까지 프로젝트를 진행하며 모두 고유한 번호를 붙였는데
현재 49번까지 진행했다. 고유 번호란 규모에 상관없이
중요하다는 의미이며, 실현되고 있거나 완공되었다는 의미도
가진다. 초기에는 소규모 프로젝트가 많았는데 대부분
지인을 통해 시작한 작업이었다. '글램핑 파빌리온Glamping
Pavilion 프로젝트' '화이트큐브 망우' '레드스퀘어하우스Red
Square House' '방배동 지층사무실' '베지가든Vegegarden' 등이
그렇다. 모든 프로젝트는 우리의 레퍼런스로 우리의 얼굴이
된다. 작은 레퍼런스가 하나둘씩 쌓이고 알려지면서 점점
프로젝트의 규모가 커졌고 건축 문의를 통한 의뢰로
프로젝트가 성사되었다. 프로젝트를 돌아보니 크게 세
가지로 분류되는 것 같다. 하나는 건축물, 다른 하나는
건축물 중에서도 특별한 구조를 가진 파빌리온, 그리고
마지막으로 예술 작업이다.
　　여기에서는 몇 가지 건축물 프로젝트를 소개하려고

#RedSquareHouse #house #whitecubeMangwoo
#BangbaedongUndergroundOffice
#ArchiWorkshopFoundationSeoulOffice

We opened the offices of ArchiWorkshop in 2013 in
a small room in the 79m² apartment where we lived.
This is how we started architecture in Korea. We
have added a serial number to each of the projects
we have taken part in up to now, and we are up to
number 49. A number means regardless of the size
of the project, all works are significant. It also
means that it is being realized or completed. In
the beginning we had many small scale projects,
and they were mostly projects which we worked on
through people we knew. The <Glamping Pavilion>,
<White Cube Mangwoo>, <Red Square House>, <Bangbae-
don Underground Office>, <Vege Garden>, all apply.
All of these projects become a point of reference
for us, the face of our firm. With each small
references accumulating one by one, and becoming
known to the world, the capacity of the projects
have gradually increased, and architectural
requests arrive more often. Looking back on
the projects, they can be categorized in three
types: Architectural projects; Pavilions, meaning
architectural projects with special structures;
Lastly art projects.
　　We would like to introduce some of our
architectural projects here. The <Red Square House>
is a house within a single unit housing lot in

↗ Red Square House
The view of the dining room.
The space is connected
directly to the ground floor
garden.

한다. ‹레드스퀘어하우스›는 경기도 부천의 전원주택 단지에
있는 주택이다. 처음 대지를 방문했을 때는 택지 개발이
거의 끝나 신축 주택이 단지를 채워가고 있었다. 대부분
비슷한 재료에 서로 닮아 있는 집들이었다. 낮게 지어진 집과
조용한 길, 듬성듬성 놓인 개인 정원을 살펴보며 이곳은
전원주택으로서의 매력이 크겠다는 생각을 했다.

　　우리는 도로를 접한 외벽만 재료를 달리 사용하면
어떨지 고민했다. 그래서 흔한 재료지만 독특해 보이도록
전면에만 빨간 벽돌을 사용했고 줄눈도 벽돌과 유사한
색으로 시공해 마치 붉은색으로 그린 커다란 그림처럼
보이게 했다. 다른 쪽 외벽은 흰색 스투코 Stucco로 마감했는데
도로에 접한 빨간 벽돌 외벽과 강렬한 대비가 연출되어
이 집만의 정체성이 형성되었다. 그리고 각 층의 공간에서
정원, 발코니, 옥상 테라스로 접근할 수 있도록 계획했다.
‹레드스퀘어하우스›는 부모님, 젊은 부부, 아이들이 함께 사는
3세대의 집이다. 부부 세대 공간과 부모님 세대 공간으로
평면을 크게 나눴지만, 1층에 서로 통하는 복도를 두어
연결고리를 만들었다. 서로의 사생활은 지켜주면서 아이들이
자유롭게 다닐 수 있도록 계획하려고 했다.

　　‹화이트큐브 망우›는 38평 대지 위에 연면적 60평

Bucheon, Gyeonggi-do. When we first visited the
site, the site development had almost come to an
end, and new houses were filling the district. They
were mostly houses which resembled each other, with
similar materials. We looked at the low set houses,
the quiet streets, the sparsely set personal
gardens, and thought that this would be a highly
attractive site for a countryside house.
　　We considered what it would be like to use
a different material just for the external wall
facing the road. Although a common material, we
distinguished the front of the house with red
bricks and used a similar color to the bricks for
the joints during the construction, so that it
would look like a big monochromatic painting. The
other outerwall was finished with white stucco, and
this choreographed a strong contrast with the red
brick tile outer wall facing the road, establishing
a unique identity for this house. All the spaces
were planned so that they could be accessed
through the garden, the balcony and the rooftop
terrace. The ‹Red Square House› is a home for
three generations, where elderly parents, a young
couple and their children all live together. While
the floor plan was largely divided between space
for the couple and the parents, a connection was
made through a mutually accessible corridor on the
first floor. This was designed so that each family
unit's privacy could be assured while allowing the
children to freely move between.

레드스퀘어하우스 (2016)
도로와 접한 입면에만 빨간 벽돌을 입혀 마치 붉은색으로
그린 커다란 그림처럼 보이도록 만들었다.

RED SQUARE HOUSE (2016)
The red bricks covered only the façade, So it
looks like a big monochromatic painting.

남짓의 규모로 지은 네 가구를 위한 다가구 주택이다.
주변이 대부분 작은 대지로 구성되어 있다 보니 건물
사이의 간격이 좁았다. 빛이 들어오지 않는 일조권 문제도
있었지만, 옆집 건물과 너무 붙어 있어 40여 년 동안 창도
열지 못한 채 살 정도로 사생활이 보호되지 않는다는 문제가
더 컸다. 우리는 내부 공간에 빛이 들어오도록 하늘로 열린
햇빛 발코니를 생각했다. 내부는 공적 공간과 사적 공간이
가족의 생활 패턴에 따라 적절하게 나뉘도록 구성했다.
또한 규모가 작더라도 설계 초기 단계부터 설비, 기계, 전기
등의 모든 배선 배관을 파이프 샤프트Pipe Shaft로 정리하고
우수관도 외부에 노출되지 않도록 고려했다. 〈화이트큐브
망우〉를 진행하며 단독 대지 안에서만 해결해야 하는 필로티
주차장과 녹지 공간의 부족 문제를 고민할 수 있었다. 만약
개발 범위를 조금 확대해 몇 곳의 대지를 함께 묶어 블록
형태로 설계가 이루어질 수 있다면 1층에 새로운 공간이나
공동의 지하 주차장, 차가 없는 보행로, 놀이 공간 및
녹지 등을 도입하고 전기 지중화 등도 충분히 검토할 수
있을 것이라는 생각이 들었다.

　　　〈방배동 지층사무실〉은 우리 내부 프로젝트였다.
첫 사무실이었던 집에서 벗어나 서울 서초구 방배동의 한

The ‹White Cube Mangwoo› is a multiple unit
home for four families, and has a total floor
area capacity of 198m². As the majority of the
surrounding lots were composed of small sites, the
buildings were set closely together. Aside from
the issue of daylight, the extreme proximity to
the building next door posed a huge privacy issue,
where the client was unable to open their window
for forty years. We conceived sunlit balcony to
allow light to enter the interior. The interior
was composed by adequately partitioning public and
private spaces. Also, despite the small scale,
from the early stages of the design process we
considered how we could organize all the wiring
and plumbing such as facilities and mechanical
electronics, etc., using pipe shafts to avoid the
storm water pipelines (gullies) being exposed on
the outside. The ‹White Cube Mangwoo› enabled us
to consider the issues of piloti car parking and
the lack of green spaces which must be resolved
in single sites. We felt that we could have amply
considered the possibility of introducing new
spaces on the first floor, a common underground
parking area, a pedestrian road without cars, a
play and green space, as well as grounding the
electronics, if the development scope had been
expanded a little for the design to take place in a
block form, with a couple of sites bound together.
　　The ‹Bangbaedong Underground Office› was our
internal project. We had decided to move out of our

화이트큐브 망우 (2014)
단독 주택가에 지어진 네 가구를 위한 다가구 주택이다.

WHITE CUBE MANGWOO (2014)
The building is a multiple housing for
four families located in the middle of the
residential area.

건축공방 연희동 사옥 (2018)
단순하고 기본적인 건축 언어를 택하고
이를 설계에 적용했다.

ARCHIWORKSHOP FOUNDATION SEOUL OFFICE (2018)
The architects adopted and applied simple,
basic architectural language.

모퉁이 건물 1층에 작업 공간을 만들기로 한 것이다. 그곳은
15평 남짓한 크기로 삼면이 도로와 접하고 있다는 것이
특징이었다. 동네 인테리어 업체가 입점해 있던 곳이었는데,
우리는 리노베이션을 하면서 사용자 입장뿐 아니라 거리에서
내부를 바라보는 행인의 시각도 함께 고려했다. 거리를
지나는 사람들이 쉽게 우리를 보고 접근할 수 있도록
열린 공간을 만들고자 했다. 이는 답답한 주택가에 숨통을
트여주고 건축가의 작업실을 공개해 보여주고자 한 시도였다.
곧 이곳은 연주회와 모임이 열리는 커뮤니티 장소이자 동네
아이들이 들어와 건축에 대한 궁금증을 묻고 답하는 공간이
되었다.

　　　2018년 가을 ‹건축공방 연희동 사옥›이 서울
서대문구에 있는 작은 산인 안산과 평행선을 이루는
연희로에 완공되었다. 사옥이 위치한 연희동은 프랜차이즈가
아닌 개성 있고 좋은 품질의 제품을 제공하는 동네 가게가
많아 독창적인 문화를 만들어내고 있었다. 우리는 소위
‘용적률 게임’을 통해 주어진 규모의 한계 안에서 단순하고
기본적인 건축 언어를 택하고 이를 설계에 적용했다. 기본을
추구하는 힘은 지속성을 키우는 작업이라고 생각한다.
전체에서 상부 입면은 알루미늄 소재의 아노다이징

first office and house, and move our nest to the
first floor of a corner in Bangbae-dong. This place
was approximately 50m², with three open sides. It
had previously been occupied by a local interior
design firm, and while carrying out the interior,
we considered not only the user's position, but
also the perspective of the passers-by looking in
from the street. We wished to make an open space
which people could easily approach while passing
by on the street. This was an attempt to provide
some respite in the stuffy residential area, and to
open up an architect's studio. This place became a
community site hosting concerts and gatherings, as
well as a parlor, where local children could come
in and explore their questions about architecture.
　　　In the fall of 2018, ‹ArchiWorkshop Foundation
Seoul Office› was completed in Yeonhui-dong,
Seodaemun-gu, Seoul. The office is located in place
brimming with a unique culture, due to the large
number of local shops, rather than franchises, with
original and high quality products. We adopted and
applied simple, basic architectural language within
the limits of the mass given through the so-called
"floor area ratio game". We used the FAR Game to
select a simple and basic architectural language
within the given capacity limits, and applied
this to the design. We believe that the pursuit
of perfecting these fundamentals form the basis
of sustainable architecture. An anodized panel
of aluminum material was applied to the entire

건축공방 연희동 사옥 (2018)
빌딩 상부 입면에는 알루미늄 소재의 아노다이징
패널을 적용했다.

ARCHIWORKSHOP FOUNDATION SEOUL OFFICE (2018)
An anodized panel of aluminum materials was
applied to the entire surface of the upper
facade.

건축공방 연희동 사옥 (2018)
내부 설계에서는 복잡함을 넘어서는 단순함을 위해 고민했다.

ARCHIWORKSHOP FOUNDATION SEOUL OFFICE (2018)
For the interior the architects pondered on
the best way to find simple and complexity
space.

건축공방 연희동 사옥 (2018)
자유롭게 목업 작업을 할 수 있는 모형 제작 공간.

ARCHIWORKSHOP FOUNDATION SEOUL OFFICE (2018)
The model making space.

패널anodized panel을 적용했다. 이 패널로 하부 재료와 대비가
생겼고 간결한 창문 패턴을 통해 비현실적으로 느껴질 만큼
정리된 입면이 만들어졌다. 사무실이 있는 하부 공간은
땅과 연결되는 공간으로서 거친 마감재로 설계해 불규칙한
수직선을 가지는 콘크리트 입면으로 구체화했다. 건물의
정면, 측면, 후면 모두 하나의 무채색 오브제와 같은 효과를
지닌다. 하부에 사용한 콘크리트 마감은 건축물 하부와
사선 제한으로 건물 상부가 잘리는 면에도 적용했다.
내부는 복잡함을 넘어서는 단순함을 고민했고 이곳에서
미니멀 라이프 스타일이 가능하길 바랐다.

　〈건축공방 연희동 사옥〉은 크게 사무실과 주거
기능으로 나뉜다. 1-3층은 사무실이고 4-6층은 주거로
계획했다. 갤러리도 있는데 이곳에서는 콘서트, 건축
작업 전시회, 오픈 스튜디오 등 다양한 문화 활동이
열린다. 사무실에서는 늘 작업 과정을 고민하고 직접 맡은
프로젝트가 아니더라도 내용을 함께 공유하는 것을 중요하게
생각한다. 기회가 있을 때마다 워크숍을 열고 아이디어를
공유하는 것도 비슷한 이유에서다.

surface of the upper facade. It contrasts with the
underlying material and displays an immaculate
façade with a concise window pattern. The building
has the same effect in the front, sides and rear,
like an achromatic object. A rough material was
used as the finish for the lower spaces where
the offices are located, as a connection to the
ground, and materialized as a concrete facade with
irregular vertical lines. This elevation was also
applied to the lower part of the entire mass and
to the upper part where the mass was cut to comply
with setback regulations. For the interior, we
wanted a minimalist lifestyle and we pondered on
the best way to simplify complexity.

　　The ‹ArchiWorkshop Foundation Seoul Office› is
largely divided into the office and the residential
function. From the first to three floors are
offices, while the fourth, fifth and sixth floors
are residential. The ArchiWorkshop Gallery hosts
a variety of cultural events, including concerts,
architectural project exhibitions, open studios,
etc. The office always thinks the working process,
and to share content even if it is not a project
one is directly involved in. It is for similar
reasons that we open workshops and share ideas
whenever possible.

# 일상, 그 이상

# The everyday
# and beyond

Essay Five

일상의 건축이 갈 길은 아직 남아 있다. 그리고 우리는 그
이상을 꿈꾸는 건축가다. 우리가 만든 건축이 형태, 재료,
개념, 그 무엇이든 우리 삶의 풍경과 정체성에 긍정적으로
작용하기를 바란다.

우리가 작업한 건축물 가운데 특별한 구조를 가진
파빌리온 프로젝트가 있다. 자연환경과 최대한 가까운
곳에 공간을 설치해 사람들에게 편안하고 안락한 쉼터를
제공하는 '글램핑 파빌리온 프로젝트'다. 텐트의 외피는
멤브레인Membrane이라는 막 구조로 설계했기 때문에
"이게 건축인가?" 하고 의심받기도 한다. 일반적인 건축물
이미지에서 벗어난 작업이기 때문이다. 콘크리트가 아니면
건축이 아니라는 선입견은 여전히 존재한다. 우리는 여가
생활과 캠핑 문화를 관심 있게 들여다보며, 레저 문화에서
디자인과 기능 그리고 무엇보다 안전성을 고려한 독특한
공간을 공유하고자 했다.

경기도 양평에 있는 첫 번째 글램핑 파빌리온 ‹양평
생각 속의 집›은 산과 강가에 있는 조약돌에서 영감을 받아

Everyday architecture still has a long way to
go. And we are architects who dream of something
beyond this. Whatever we make, whether it be form,
material, or concept, we hope that what we create
can have a positive impact on the landscapes and
identities of our lives.

Among the projects we have worked on, are
some pavilion projects with special structures.
By installing a space at the closest point to
the natural environment, we created one glamping
pavilion which provides a comfortable and cozy
space for people. Some doubt whether this project
can be considered architecture, as the outer-skin
of the tent is designed with a canvas structure
called a membrane. It is a project which deviates
from what we think of as a normal design method.
The prejudice that non-concrete structures are not
considered as architecture still exists. We have
attempted to take a close look at leisure-based
lifestyles and camping culture, and to share unique
spaces which consider design, function, and above
all, safety, in leisure culture.

The ‹Yangpyeong Mind Home›, The first glamping
pavilion in Yangpyeong, Gyeonggi-do was designed by
utilizing the method of stacking circular forms,
inspired by the pebbles in the mountains and the
river. It was created by considering the DIY

양평 생각 속의 집 (2013)
자연환경과 최대한 가까운 곳에 공간을 설치해 편안하고
안락한 쉼터를 제공하는 글램핑 파빌리온이다.

YANGPYEONG MIND HOME (2013)
It is the glamping pavilions that provides
a space for rest and relaxation as close as
possible to the natural environment.

가평바위숲 '온더락' (2015)
건축물은 파빌리온 영역이 온전히 보호될 수 있도록
숲과 도로의 경계 역할을 한다.

ROCK FOREST 'GLAMPING ON THE ROCK' (2015)
The architecture building could act as a
boundary between forest and road protecting
the pavilion zone.

가평바위숲 '온더락' (2015)
(위) 파빌리온의 배치는 자연과 만나는 태도를 고민한 결과다.
(아래) 내부의 건축 기둥은 대지에 있던 잣나무를 활용했다.

ROCK FOREST 'GLAMPING ON THE ROCK' (2015)
(top) Considering about the way of meeting
the nature was important.
(bottom) The internal column of building is
the pine trees on the ground.

원을 쌓아 올리는 방식을 적용해 설계했다. 가변적 작업을
기초로 한 조립식 시공법을 고려해 만들었다. 다양한
장소에서 자연의 훼손을 최소화하고 사용자가 자연과 건축을
경험하며 머무를 수 있도록 했다.

　　자연 속에 지은 또 하나의 글램핑 파빌리온
프로젝트는 바위가 많은 잣나무 숲속에 자리 잡은 ‹가평
바위숲 ‘온더락’›이다. 잣나무숲 끝자락에 자리 잡은 대지는
북측 끝에서 남측 끝까지 높이가 25미터 차이가 났다.
파빌리온 영역이 온전히 보호될 수 있도록 건축물이 숲과
도로의 경계 역할을 하도록 구상했다. 건축물의 경계와
경사를 따라 세 개의 덩어리로 나누어 배치하고 그 중 공공
영역으로 쓰일 마지막 건축물은 숲의 모습을 담을 수 있도록
차분한 색조의 미러 글라스와 목재로 마감했다. 건물 내부
기둥은 대지에 있던 잣나무를 그대로 활용했다.

　　‹캠프통 아일랜드 & 미술관›은 경기도 가평군
청평호수 근처에 위치한 복합 리조트다. 삼각형 모양의
대지에 처음 방문했을 때 마치 호수 위에 떠 있는 듯한
느낌을 받았다. 뱃머리에 서 있는 기분이었다. 전체 대지가
편마암으로 이루어져 있어 땅의 이미지는 단단하고 견고했다.
우리는 독특한 체험을 할 수 있는 레저 공간을 위해 땅에서

construction methods used in flexible projects.
It was designed to minimize damage to nature in a
variety of locations, so that the user could stay
while experiencing nature and architecture.
　　Another glamping pavilion project built
within nature is the <Rock Forest 'Glamping On The
Rock'> located deep within a rocky pine forest.
The site which is situated at the end point of a
pine forest has an elevation difference of 25m
from the northern most end point to the southern-
most end point. It was composed so that the extent
of the pavilion could be completely projected,
and the architecture building could act as a
boundary between forest and road. The architecture
was arranged in three masses according to the
boundaries and slopes, and among these, the final
architectural project, used in the public spaces,
was finished with toned down mirrored glass and
wood, embodying the scenes of the forest. The
columns on the interior of the building used the
pine trees found at the same location.
　　The <Camptong Island & Museum> is a resort
complex located nearby the Chungpyung Lake in
Gapyeong-gun, Gyeonggi-do. When we visited the
triangular site, we felt as if it were floating
on top of the lake, or like we were standing at
the helm of a boat. The entire site is made up of
gneiss, composing a firm and sturdy image of the
land. We started with a masterplan inspired by the
land, and created a leisure space where people

캠프통 아일랜드 & 미술관 (2017)
대지에 있던 병풍바위는 호수를 조망하는 플랫폼,
즉 열린 광장으로 재해석되었다.

CAMPTONG ISLAND & MUSEUM (2017)
The rock on the site was regarded as an open
plaza, a platform overlooking the lake.

받은 영감으로 마스터 플랜을 시작했다. 대지 규모는 약
2,500평으로 크게 외부 공간, 미술관, 레스토랑, 웰컴센터,
글램핑 공간으로 구분했고, 누구나 방문이 가능한 공공
영역, 미술관과 같은 준 공공 영역, 숙박을 할 수 있는
글램핑 구역의 개인 영역으로 구분해 출입 동선을 계획했다.
그곳에 있는 병풍바위는 호수를 조망하는 플랫폼 즉 열린
광장으로 재해석되었다. 필요한 공간들은 바위 속에 박힌
장소로 구현되어 건축이 드러나지 않고 땅의 영역 안에서
만들어져 건물의 지붕이 플랫폼이 되거나 광장이 되도록
설계했다. 글램핑 파빌리온을 광장보다 낮은 높이에 배치해
열린 광장에서 청평호수가 보이도록 했다. 병풍바위와
웰컴센터의 건축적 구조를 활용해 도로에서 들려오는 소음을
막고 보호받는 공간으로 만들었으며 상부 전망 플랫폼으로
연결하는 계단을 만들었다.

우리는 다양한 예술 설치 작업도 진행한다. 이런
설치 작업이 재미있는 이유는 공간적 상상력을 자극하고
그것이 다시 건축 작업에 영감을 주기 때문이다. 건축과
예술 작업을 통해 재료를 개발하고 재활용 제품을 만들기도
한다. 서울 서대문구 북아현동 재개발 지역에 있던 철거되기
바로 직전의 다세대 주택에 설치한 ‹삶의 환영, Welcome

could have a unique experience. The site capacity was approximately 8.260m², divided largely into the outdoor space, the museum, the restaurant, the welcome center, the glamping spaces, while circulating pathways were planned by dividing these into the public spaces, where anyone could visit, the semi-public spaces such as the museum, and finally the private spaces of the glamping area where people could reside. The rocks here were reinterpreted as a platform overlooking the lake, and hence an open plaza. The necessary spaces were realized as places embedded in the rocks, formed in harmony with the land rather than revealing themselves as architecture, and the roofs of the building were designed to become platforms or plazas. By arranging the glamping pavilions at a lower elevation than the plaza, it was possible to see Chungpyung Lake from the open plaza. By utilizing the architectural structure of the folding screen *byeong-pung* rock and the welcome center, it was possible to block out the noise from the road and create a protected space, while simultaneously forming a set of stairs which connect to the upper level scenic viewpoint platform.

We have also worked on a variety of art installation projects. These installation projects are fun because they stimulate our spatial imagination and this in turn serves as an inspiration to our architectural projects. We also

삶의 환영, Welcome 그리고 Illusion (2017)
건축공방과 여러 예술 작가가 협업하여 진행한
설치 작업이었다. '환영'과 같은 공간을 조성했다.

WELCOME AND ILLUSION (2017)
It was an installation artwork that was
collaborated with the ArchiWorkshop and
various installation artists. It created a
space like 'illusion'.

그리고 Illusion›은 서울문화재단의 후원으로 완성되었다.
이 작업은 건축공방과 예술 작가 이창훈, 사운드 작가 김준,
바이올리니스트 심혜선, 첼리스트 심혜원과의 협업으로
진행했다. 다원 예술의 성격으로, 빠르게 변화하는 우리
도시의 모습을 돌아보며 공유하고 기억하고자 한 기록
작업이다. 재개발로 비워진 집에 '환영 illusion'과 같은
공간 설치를 진행했고, 이 작업을 통해 새로운 장소를
'환영 welcome'하는 곳이 되도록 의도했다. 약 2년 동안의 준비
기간을 거쳐 진행한 작업으로 우리 주변에서 일어나는
현상에 관심을 가지고 고민해 나온 작품이었다. 한국의
도시 재개발 상황과 이를 바라보는 건축가와 작가들의
시선을 보여주고 공간적으로 구현했다.

　　　한강예술공원에 설치했던 파일럿 프로젝트
‹바다바람›은 폐어선을 한강공원에 가져다 놓고 바람의
움직임과 소리를 통해 바다와 그늘을 느낄 수 있도록 한
설치 작업이다. 이 프로젝트를 통해 예술과 일상이 더 가깝게
마주해 시민에게 개방되기를 바랐다.

　　　우리가 진행한 모든 비건축적 작업은 확장된 건축
작업이다. '일상의 건축'에서 뜻하는 일상은 지루하게
반복되는 생활이 아니라 어디서나 마주치게 되는 공간이다.

develop materials and create recycled products
through our architectural and art projects. The
‹Welcome and illusion› was installed in the
multiple housing lot just before the Buk-Ahyun-
dong, Seodaemun-gu, Seoul redevelopment district
was demolished and launched with the support of the
Seoul Cultural Foundation.
　　　This project took place through the
collaboration of ArchiWorkshop, the installation
artists Changhoon Lee, sound artists Jun Kim,
the violinist Hyesun Sim, and the cellist Hyewon
Sim. Characterized as a multidisciplinary art, it
was an archiving project which attempted to look
back on, and share and remember what our rapidly
transforming cities once looked like. A spatial
installation, such as "illusion" was inserted into
a house emptied out due to redevelopment, and this
project was intended to "welcome" the new place.
The circumstances of urban redevelopment in Korea,
and the perspectives of the architects and artists
observing this were featured and realized through
space.
　　　The pilot project <Blue Wind> set at the
Hangang Arts Park was an installation work to feel
the sea and the shadows through the movement and
sound of the wind by placing an abandoned fishing
boat. We hoped that it could be opened to the
public for closer contact between art and daily
life.
　　　All of the non-architectural works we have

바다바람 (2017)
폐어선을 한강공원에 가져다 놓고 바람과 소리를 통해
바다와 그늘을 느낄 수 있도록 한 설치 작업이다.

BLUE WIND (2017)
It is installation artwork to feel the sea
and the shadows through windy and sound by
placing an abandoned fishing boat.

가평바위숲 '온더락' (2015)
외피의 막구조 덕분에 사용자는 색다른 공간감을
느낄 수 있다.

ROCK FOREST 'GLAMPING ON THE ROCK' (2015)
The users could feel a unique atmosphere at
internal space due to the membrane structure.

일상적인 건축의 가치를 높이는 것, 그것이 좋은 건축이라고
생각한다. 누구나 인간적이고 사회적이고 친밀하고 열려
있고 접근 가능하고 민주적이고 자유롭고 다양하고 역사적인
것을 추구한다. 그리고 '아름답고' '사랑스럽고' '흥미진진한'
같은 추상적인 단어들을 말한다. 그러나 '어떻게 만들
것인가?'라는 질문은 여전히 남아 있다. '어떻게'는 고민의
시작이고 실험을 통한 과정이고 건축의 결과다.

우리는 기본에서 시작한다. '건축공방'이라는 이름이
가지는 '공예가의 작업실 Workshop'이라는 의미와 '서로
공격하고 방어하는 토론'이라는 의미를 바탕에 두고
실천하며, 우리의 일상성이 특별해지는 건축, 그런 슈퍼
하드웨어를 공유한다. 건축은 동굴이나 쉘터와 같은 가장
기초적인 부분만이 아니라 문화와 생각을 만들어내는
영향력을 가지고 있기 때문이다. "사람이 책을 만들고, 책이
사람을 만든다."는 말이 있듯이, 사람이 건축을 만들고
건축(환경)이 사람을 만든다. 작은 힘들이 모여, 방향성을
만들 듯이 이로운 건축이 모이고 알려지고 공유되어 새로운
방향성을 만든다. 우리 시대의 건축과 문화가 미래의
유산으로 남을 수 있는 길을 만든다.

carried out are extended architectural works.
The daily evoked through "everyday architecture"
is a space that one can come across anywhere,
rather than a boringly repetitive lifestyle. We
believe that good architecture involves increasing
the value of everyday architecture. Everybody
pursues that which is humane, social, intimate,
open, accessible, democratic, liberal, diverse
and historical. And they also say abstract terms
like 'beautiful' 'lovely' 'exciting.' However,
the question of 'how this might be made' always
remains. "How" is the starting point of this task,
and is a process through experimentation, and a
result of the architecture.

We start with the basics. Based on the
meaning ArchiWorkshop came from the 'workshop for
the craftsman' and 'discussion', we share the
architecture and super hardware that make our daily
life special. This is because architecture has
the power to create culture and thoughts as well
as taking on functional basic roles like caves
or shelter. Like the saying "People make books
and books make people," people make architecture
and architecture (and environments) makes people.
Small efforts come together to create a sense
of direction, in the same way that beneficial
architecture is gathered, made known and shared,
and establishes a new sense of direction. The
architecture and culture of our age will create a
road which can remain as a legacy for the future.

↗ From the left, Sujeong Park
and Heejun Sim, principals of
ArchiWorkshop.

설계 개요

양평 생각 속의 집

- 설계 ¦ 건축공방(심희준, 박수정)
- 위치 ¦ 경기도 양평
- 용도 ¦ 리조트
- 대지면적 ¦ 3,300m²
- 건축면적 ¦ 346.5m²
- 연면적 ¦ 346.5m²
- 규모 ¦ 지상 1층
- 높이 ¦ 3.5m
- 주차 ¦ 7대
- 건폐율 ¦ 10%
- 용적률 ¦ 10%
- 구조 ¦ 스틸프레임
- 외부마감 ¦ 2중 멤브레인(백색)
- 내부마감 ¦ 멤브레인, 목재
- 구조설계 ¦ 건축공방
- 기계·전기설계 ¦ (주)선진설비콘설탄트
- 시공 ¦ 건축공방
- 설계기간 ¦ 2013. 8. - 9.
- 시공기간 ¦ 2013. 10. - 12.
- 사진 ¦ 임준영
- 건축주 ¦ 개인

Design overview

YANGPYEONG MIND HOME

- Architect ¦ ArchiWorkshop (Heejun Sim, Sujeong Park)
- Location ¦ Yangpyeong-gun, Gyeonggi-do, Korea
- Programme ¦ resort
- Site area ¦ 3,300m²
- Building area ¦ 346.5m²
- Gross floor area ¦ 346.5m²
- Building scope ¦ 1F
- Height ¦ 3.5m
- Parking capacity ¦ 7
- Building coverage ¦ 10%
- Floor area ratio ¦ 10%
- Structure ¦ steel frame
- Exterior finishing ¦ 0000
- Interior finishing ¦ 0000
- Structural engineer ¦ ArchiWorkshop
- Mechanical and electrical engineer ¦ Sunjin Engineering Consultant Co.,Ltd.
- Construction ¦ ArchiWorkshop
- Design period ¦ Aug. - Sep. 2013
- Construction period ¦ Oct. - Dec. 2013
- Photograph ¦ Juneyoung Lim
- Client ¦ Private

화이트큐브 망우

- 설계 ┆ 건축공방(심희준, 박수정)
- 위치 ┆ 서울시 중랑구 망우동
- 용도 ┆ 주택
- 대지면적 ┆ 125.4m²
- 건축면적 ┆ 72.6m²
- 연면적 ┆ 191.4m²
- 규모 ┆ 지상 4층
- 높이 ┆ 11.5m
- 주차 ┆ 3대
- 건폐율 ┆ 60%
- 용적률 ┆ 158.5%
- 구조 ┆ 철근콘크리트조
- 외부마감 ┆ 외부단열 위 스타코플렉스(백색, 흑색)
- 내부마감 ┆ 석고보드 위 벽지
- 구조설계 ┆ 계명구조
- 기계·전기설계 ┆ 진경 ENG
- 시공 ┆ 공정건설(주)
- 설계기간 ┆ 2014. 1. - 5.
- 시공기간 ┆ 2014. 6. - 11.
- 사진 ┆ 임준영
- 건축주 ┆ 개인

가평바위숲 '온더락'

- 설계 ┆ 건축공방(심희준, 박수정)
- 위치 ┆ 경기도 가평군
- 용도 ┆ 주택, 웰컴센터, 카페, 글램핑
- 대지면적 ┆ 2,300m²
- 건축면적 ┆ 222.4m²
- 연면적 ┆ 268.3m²(글램핑 리조트시설 346.5m²)
- 규모 ┆ 2층
- 높이 ┆ 4.4m(최대)
- 주차 ┆ 3대
- 건폐율 ┆ 10%
- 용적률 ┆ 10%
- 구조 ┆ 철근콘크리트조, 스틸기둥, 경량목구조
- 외부마감 ┆ 시더우드, 미러글라스, 스타코플렉스
- 내부마감 ┆ 벽지, 강마루
- 구조설계 ┆ TCM Global
- 기계·전기설계 ┆ TCM Global
- 시공 ┆ TCM글로벌
- 설계기간 ┆ 2014. 3. - 6.
- 시공기간 ┆ 2014. 7. - 2015. 3.
- 사진 ┆ 임준영
- 건축주 ┆ 온더락

WHITE CUBE MANGWOO

- Architect ┆ ArchiWorkshop (Heejun Sim, Sujeong Park)
- Location ┆ Mangwoo-dong, Jungnang-gu, Seoul, Korea
- Programme ┆ housing
- Site area ┆ 125.4m²
- Building area ┆ 72.6m²
- Gross floor area ┆ 191.4m²
- Building scope ┆ 4F
- Height ┆ 11.5m
- Parking capacity ┆ 3
- Building coverage ┆ 60%
- Floor area ratio ┆ 158.5%
- Structure ┆ RC
- Exterior finishing ┆ stuco-flex, external insulation (White/Black)
- Interior finishing ┆ wall paper on gypsum boards
- Structural engineer ┆ Kei Myung Structural Engineering & Consulting co., Ltd
- Mechanical and electrical engineer ┆ Jinkyung ENG
- Construction ┆ Fair Design Construction
- Design period ┆ Jan. - May 2014
- Construction period ┆ June - Nov. 2014
- Photograph ┆ Juneyoung Lim
- Client ┆ Private

ROCK FOREST 'GLAMPING ON THE ROCK'

- Architect ┆ ArchiWorkshop (Heejun Sim, Sujeong Park)
- Location ┆ Gapyeong-gun, Gyeonggi-do, Korea
- Programme ┆ housing, welcome center, cafe, glamping
- Site area ┆ 2,300m²
- Building area ┆ 222.4m²
- Gross floor area ┆ 268.3m²(Glamping resort 346.5m²)
- Building scope ┆ 2F
- Height ┆ 4.4m (max.)
- Parking capacity ┆ 3
- Building coverage ┆ 10%
- Floor area ratio ┆ 10%
- Structure ┆ RC, steel column, light wood structure
- Exterior finishing ┆ cedarwood, mirror glass, stuco-flex
- Interior finishing ┆ wall paper, wood
- Structural engineer ┆ TCM Global
- Mechanical and electrical engineer ┆ TCM Global
- Construction ┆ TCM Global
- Design period ┆ Mar. - June 2014
- Construction period ┆ July 2014 - Mar. 2015
- Photograph ┆ Juneyoung Lim
- Client ┆ On the Rock

레드스퀘어하우스
- 설계 ｜ 건축공방(심희준, 박수정)
- 위치 ｜ 경기도 부천시 오정구 작동
- 용도 ｜ 주택
- 대지면적 ｜ 218m²
- 건축면적 ｜ 107.31m²
- 연면적 ｜ 211.11m²
- 규모 ｜ 지상 3층
- 높이 ｜ 9m
- 주차 ｜ 2대
- 건폐율 ｜ 49.22%
- 용적률 ｜ 96.84%
- 구조 ｜ 철근콘크리트
- 외부마감 ｜ 적벽돌타일, 스타코플렉스
- 내부마감 ｜ 석고보드 위 벽지
- 구조설계 ｜ (주)계명구조엔지니어링
- 기계·전기설계 ｜ (주)진경엔지니어링
- 시공 ｜ 건축공방
- 설계기간 ｜ 2015. 10. - 2016. 2.
- 시공기간 ｜ 2016. 3. - 10.
- 사진 ｜ 이남선
- 건축주 ｜ 개인

하늘을 담은 집
- 설계 ｜ 건축공방(심희준, 박수정)
- 위치 ｜ 서울시 중랑구 망우동
- 용도 ｜ 주택
- 대지면적 ｜ 108.9m²
- 건축면적 ｜ 53.7m²
- 연면적 ｜ 172.6m²
- 규모 ｜ 지상 5층
- 높이 ｜ 14.2m
- 주차 ｜ 2대
- 건폐율 ｜ 49.27%
- 용적률 ｜ 158.27%
- 구조 ｜ 철근콘크리트조
- 외부마감 ｜ 외부단열 위 검은색 벽돌타일, 스타코플렉스, 블랙 미러 패널
- 내부마감 ｜ 석고보드 위 벽지
- 구조설계 ｜ G&H 디자인 워크숍
- 기계·전기설계 ｜ 진경 ENG
- 시공 ｜ 건축공방
- 설계기간 ｜ 2016. 2. - 5.
- 시공기간 ｜ 2016. 7. - 12.
- 사진 ｜ 임준영
- 건축주 ｜ 개인

RED SQUARE HOUSE
- Architect ｜ ArchiWorkshop (Heejun Sim, Sujeong Park)
- Location ｜ Jak-dong, Ojeong-gu, Bucheon, Gyeonggi-do, Korea
- Programme ｜ housing
- Site area ｜ 218m²
- Building area ｜ 107.31m²
- Gross floor area ｜ 211.11m²
- Building scope ｜ 3F
- Height ｜ 9M
- Parking capacity ｜ 2
- Building coverage ｜ 49.22%
- Floor area ratio ｜ 96.84%
- Structure ｜ RC
- Exterior finishing ｜ BRICK TILE, STUCOFLEX
- Interior finishing ｜ PAPER ON THE BOARD
- Structural engineer ｜ Kyemyung Structure Engineering
- Mechanical and electrical engineer ｜ Jinkyeong Engineering
- Construction ｜ ArchiWorkshop
- Design period ｜ Oct. 2015 - Feb. 2016
- Construction period ｜ Mar. - Oct. 2016
- Photograph ｜ Namsun Lee
- Client ｜ private

HOUSE EMBRACING SKY
- Architect ｜ ArchiWorkshop (Heejun Sim, Sujeong Park)
- Location ｜ Mangwoo-dong, Jungnang-gu, Seoul, Korea
- Programme ｜ housing
- Site area ｜ 108.9m²
- Building area ｜ 53.7m²
- Gross floor area ｜ 172.6m²
- Building scope ｜ 5F
- Height ｜ 14.2m
- Parking capacity ｜ 2
- Building coverage ｜ 49.27%
- Floor area ratio ｜ 158.27%
- Structure ｜ RC
- Exterior finishing ｜ brick tile, stucoflex, external insulation, black mirror panel
- Interior finishing ｜ wall paper on gypsum boards
- Structural engineer ｜ G&H Design Workshop
- Mechanical and electrical engineer ｜ Jinkyeong ENG
- Construction ｜ ArchiWorkshop
- Design period ｜ Feb. - May 2016
- Construction period ｜ July - Dec. 2016
- Photograph ｜ Juneyoung Lim
- Client ｜ Private

캠프통 아일랜드 & 미술관
- 설계 ┊ 건축공방(심희준, 박수정)
- 위치 ┊ 경기도 가평군 청평면
- 용도 ┊ 글램핑 리조트, 미술관, 웰컴센터, 레스토랑
- 대지면적 ┊ 7,748m²
- 건축면적 ┊ 986.57m²
- 연면적 ┊ 1,201.83m²
- 규모 ┊ 지상 1층, 지하 2층
- 높이 ┊ 4.5m
- 주차 ┊ 18대
- 건폐율 ┊ 12.73%
- 용적률 ┊ 24.31%
- 구조 ┊ 철근콘크리트조, 스틸멤브레인(글램핑 파빌리온)
- 외부마감 ┊ 노출콘크리트, 2중 멤브레인(글램핑 파빌리온)
- 내부마감 ┊ 페인트
- 구조설계 ┊ (주)계명구조엔지니어링
- 기계·전기설계 ┊ (주)선진설비콘설탄트
- 시공 ┊ 건축공방(글램핑 파빌리온)
- 설계기간 ┊ 2014. 11. - 2015. 11.
- 시공기간 ┊ 2016. 3. - 2017. 6.
- 사진 ┊ 정정호
- 건축주 ┊ (주)캠프통아일랜드

건축공방 연희동 사옥
- 설계 ┊ 건축공방(심희준, 박수정)
- 위치 ┊ 서울시 서대문구 연희로 193-8
- 용도 ┊ 업무시설, 주거
- 대지면적 ┊ 304.6m²
- 건축면적 ┊ 150m²
- 연면적 ┊ 756.46m²
- 규모 ┊ 지상 6층
- 높이 ┊ 19m
- 주차 ┊ 7대
- 건폐율 ┊ 59.95%
- 용적률 ┊ 198.95%
- 구조 ┊ RC
- 외부마감 ┊ 아노다이징패널, 콘크리트 줄눈 마감
- 내부마감 ┊ 롱타일, 석고보드 위 페인트
- 구조설계 ┊ 황경주(서울시립대 건축학과)
- 기계·전기설계 ┊ (주)선진설비콘설탄트
- 시공 ┊ (주)이재605(최호근)
- 설계기간 ┊ 2016. 12. - 2017. 5.
- 시공기간 ┊ 2017. 8. - 2018. 7.
- 사진 ┊ 정정호
- 건축주 ┊ 건축공방

CAMPTONG ISLAND & MUSEUM
- Architect ┊ ArchiWorkshop (Heejun Sim, Sujeong Park)
- Location ┊ Cheongpyeong-myeon, Gapyeong-gun, Gyeonggi-do, Korea
- Programme ┊ resort, museum, welcome center, restaurant
- Site area ┊ 7,748m²
- Building area ┊ 986.57m²
- Gross floor area ┊ 1,201.83m²
- Building scope ┊ B2, 1F
- Height ┊ 4.5m
- Parking capacity ┊ 18
- Building coverage ┊ 21.73%
- Floor area ratio ┊ 24.31%
- Structure ┊ RC, steel & membrane (glamping pavilion)
- Exterior finishing ┊ exposed concrete, stacking stones
- Interior finishing ┊ painting
- Structural engineer ┊ Kyemyung Structure Engineering
- Mechanical and electrical engineer ┊ Sunjin Engineering Consultant Co.,Ltd.
- Construction ┊ ArchiWorkshop (glamping pavilion)
- Design period ┊ Nov. 2014 - Nov. 2015
- Construction period ┊ Mar. 2016 - June 2017
- Photograph ┊ Jungho Jung
- Client ┊ Camptong Island

ARCHIWORKSHOP FOUNDATION SEOUL OFFICE
- Architect ┊ ArchiWorkshop (Heejun Sim, Sujeong Park)
- Location ┊ 193-8, Yeonhui-ro, Seodaemun-gu, Seoul, Korea
- Programme ┊ office, housing
- Site area ┊ 304.6m²
- Building area ┊ 150m²
- Gross floor area ┊ 756.46m²
- Building scope ┊ 6F
- Height ┊ 19M
- Parking capacity ┊ 7
- Building coverage ┊ 59.95%
- Floor area ratio ┊ 198.95%
- Structure ┊ RC
- Exterior finishing ┊ ANODIZING PANEL,
- Interior finishing ┊ LONG TILE, PAINTING ON THE BOARD
- Structural engineer ┊ Kyungju Hwang (University of Seoul, Department of Architecture)
- Mechanical and electrical engineer ┊ Sunjin Engineering Consultant Co.,Ltd.
- Construction ┊ IJAE605 (Hogeun Choi)
- Design period ┊ Dec. 2016 - May 2017
- Construction period ┊ Aug. 2017 - July 2018
- Photograph ┊ Jungho Jung
- Client ┊ ArchiWorkshop

크리틱

# 일상과 사물의 정착
: 조남호



Critique

# Establishing the Everyday and Its Objects
: Namho Cho

조남호
조남호는 서울시립대학교 건축학부를
졸업하고, 성균관대학교 디자인대학원
건축도시디자인학과에서 석사를
마쳤다. 현재 솔토지빈건축사사무소
대표이며 서울시 건축정책위원이다.

# 일상의 건축

일상성이라는 학문적 개념은 독일의
실존주의 철학자 마르틴 하이데거Martin
Heidegger가 정립했다. 하이데거는 자신의
철학서『존재와 시간 Sein Und Zeit』에서 인간의
일상성을 다양한 층위에서 설명한다. 그는
인간이 처한 실존의 구조를 '현존재와 세계-
내-존재'로서 분석한다. 현존재가 제일 먼저
만나는 것은 자신이며, 존재의 틀을 구성하는
세계-내-존재가 현존재의 일상적 조건이다.
현존재는 주변에 대해 무차별적으로
열려 있는데, 이것을 하이데거는 평균성이라
부른다. 현상학은 개인이 현존재의 평균적
일상성 속에서 어떻게 존재하는지 밝혀내는
것이다. 현존재의 세계는 공동 세계이며
그 '안에-있음'은 타인과 더불어 있는 것이다.
보통 우리가 공공성 또는 군중이라는 말로
규정하는 그들이 일상성의 존재 양식을
지정한다. '그들'이 타당하게 여기는 것과

Namho Cho
Namho Cho studied architecture
at University of Seoul and
gained his M.A. at the
Department of Architectural
and Urban Space at the Graduate
School of Design, Sungkyunkwan
University He is the principal
of Soltozibin Architects and
currently serves on the Seoul
Architectural Policy Committee.

## EVERYDAY ARCHITECTURE

The academic concept of everydayness
(Alltäglichkeit) was established by
the German existentialist philosopher
Martin Heidegger. In 『Sein Und Zeit』,
Heidegger explains the everydayness
of men on a variety of levels. He
analyzes the existential structure of
mankind as 'Being-with (Dasein) and
Being-in-the-world.' Dasein is the
inclination to take up a relationship
towards the world, and Heidegger calls
this a levelling or averageness.
Phenomenology is the task of
uncovering how an individual exists
in the average everyday of the Being
-with. The world of the Being-with is
a common world, denoting a state of
"being-lost in the publicness of the
'they'. It is generally those who we
define as publicness or the masses
that designate the way in which the
everyday exists. What 'they' consider
to be viable or not, what they
recognize or reject, what they accept
as executable or not, establishes a
levelling of the potential of all
being. As such, the everyday Being-
with responds to the world by facing

그렇지 않게 여기는 것, 인정하는 것과 거부하는 것, 감행해도 되는 것과 안 되는 것 등은 모든 존재 가능성의 평준화를 만들어내기도 한다. 이처럼 일상의 현존재는 다양한 층위의 현상에 직면하면서 세계에 반응한다. 일상성은 현존재가 그 안에 머무는 실존의 양식을 의미하고, 평생 현존재를 두루 지배하는 실존의 특정한 방식을 의미한다. 즉 현존재가 익명의 주변 사람에게 자신을 내보이는 방식이다.

> "건축공방의 건축 철학, 그 중심에 '일상의 건축'이 있다. 일상적인 건축의 가치를 높이는 것, 그것이 건축공방이 생각하는 좋은 건축이다. 일상이라는 친숙함 때문에 '일상의 건축'이나 '좋은 건축'은 뻔하거나 누구나 다 알고 있는 내용으로 인지되기 마련이다. 그럼에도 우리가 여전히 관심을 갖는 이유는 기본에

대해 이야기하고 싶고, 기본에서부터 출발하고 싶기 때문이다."[1]

하이데거는 일상생활에 대해 따분하고 기계적으로 반복하는 하찮은 삶으로 규정하고 진정한 삶과 대립하는 개념으로 이해했다. 프랑스 철학자 앙리 르페브르H. Lefebvre는 일상의 장소는 본질적으로 모순을 안고 있고, 일상은 안정적인 듯하면서도 과정적이며 불완전하다고 인식한다. 일상의 장소는 안정을 지향하는 동시에 사회적 실천과 재생산의 시공간으로 혁신이 시도되는 장이다.

건축가 박수정, 심희준이 이끄는 건축공방의 일상에 대한 인식은 르페브르의 입장에 가깝다. 그들은 건축 프로젝트를 진행할 때 외부에서 이식된 주제를 바탕으로 하기보다는 일상에 존재하는 모순과 불균형에서 변화의 동기를 끌어낸다. 건축공방의 관심은 "우리 도시와 건축의

phenomena on a variety of levels. The everyday-ness signifies a way of existing in which the Being-with resides within, signifying the particular way of existence with which the Being-with is overall manipulated with. Hence, the Being-with is the method with which the self exposes itself to its anonymous neighbors.

> "At the core of ArchiWorkshop's architectural philosophy is the concept of "everyday architecture." Adding value to everyday architecture is ArchiWorkshop's definition of good architecture. It is due to the familiarity of the everyday that everyday architecture or good architecture is often perceived as self-evident or something that everyone already knows. Nevertheless, we remain interested in this due to our need to discuss, or perhaps one should say, to start from the basics."[1]

Heidegger viewed the everyday as the

trivial and mundane life of mechanical repetition, understanding this as a concept in conflict with authentic life. The French philosopher Henri Lefebvre perceived that places in the everyday are inherently driven by contradiction, and that the everyday may seem stable but is part of a process, and therefore incomplete. The places of the everyday are a realm of attempted innovation in the time-space of social practice and reproduction while simultaneously promoting stability.

Led by Sujeong Park and Heejun Sim, ArchiWorkshop's take on the perception of the everyday is closer to Lefebvre's position. When taking part in architectural projects, they underscore how change is motivated through the contradictions and imbalance existing in the everyday, rather than deriving this from externally transplanted themes. The interest of ArchiWorkshop connects with the architect Guyon Chung, who said "The issues of our cities and architecture exist within us, as do their solutions."

문제는 우리 안에 있고, 결국 해결책도 우리 안에 있다."는 건축가 정기용의 말에 닿아 있다.

## 일상과 사물의 정착

일상의 다양한 행위로 만들어지는 장소를 물리적으로 정착하려는 시도는 도시 건축에서 중요하다. 도시가 복잡해지고 일상의 행위들 또한 이전과는 비교가 어려울 정도로 다양해짐에 따라 이를 예측하기 위한 시도 또한 필요하다. 건축은 본래 방대한 지식과 견해와 판단으로 둘러싸여 있어 체계적으로 다루기 어렵다. 다른 한편으로 독창성이라는 이름으로 자유롭게 건축 작업을 하려는 태도가 자칫 건축을 현실과 무관한 사변으로 바라보게 할 우려가 있다. 영국 건축사학자 케네스 프램튼Kenneth Frampton은 「축조 문화에 관한 연구Studies in Tectonic Culture」에서 구축성은 건축만이 가지는 고유한 성질이며, 구축성을 통해서만이 외부적인 담론은 건축에서 실현될 수 있다고 한다. 그리고 건축이 대지site와 유형type, 구축성tectonic 간의 상호 작용 결과라고 한다면, 대지와 유형은 다양한 담론에서 현대적 경향을 가져왔지만, 구축성은 도구적 수단에 머물러왔다고 말한다. 일상에서 추출한 개념은 명확하게 구축한 작업을 통해 실현할 수 있다. 건축공방은 물리적으로 구체화한 정착 과정을 통해 일상 속 생활 환경을 좀 더 나은 곳으로 이끌면서 그 영역을 확장해간다.

## 건축의 표면

"건축공방은 일상성이 특별해지는 슈퍼하드웨어Super-hardware를 공유하는 작업을 만들고 있다. 우리가 말하는 일상의 건축은 높은 수준, 즉 건강한 환경의 건축을 지향하고, 더 많은

## ESTABLISHING THE EVERYDAY AND ITS OBJECTS

Such an attempt to physically settle places composed of diverse everyday actions is an important part of urban architecture. Such an attempt should also try to reliably predict how the city has become incomparably more complex, while everyday actions are now incomparably more diverse. Architecture is fundamentally surrounded by a vast foundation of knowledge, opinions and judgements, and hence is difficult to treat systematically. On the other hand, there is the danger that such a liberal approach to architectural design in the name of a unique sense of style will lead us to regard architecture as a matter distanced from reality. The British architectural historian Kenneth Frampton, in 「Studies in Tectonic Culture 」 defines techne- as a unique characteristic of architecture, and that external discourse can only be realized in architecture through techne-. If we were to say that architecture is the result of the interrelationship of the site, type and construction, the site and type brings forth a contemporary tendency for diverse discourses, while the tectonic has remained as a tool. Concepts derived from the everyday can be realized with clearly constructed projects. ArchiWorkshop expands its realm by leading the living environment in the everyday towards a better place through a physically articulated process of establishment.

## THE ARCHITECTURAL SURFACE

"ArchiWorkshop have created projects by sharing the Super-Hardware which render the everyday more special. The everyday architecture we speak of promotes a high quality of architecture, the architecture of a healthy environment, and an extended architectural culture that can be shared with a greater number of people."[2]

Steady contemplation of ArchiWorkshop's projects, brings a

사람이 이를 누릴 수 있도록 확장된 건축 문화를 말한다."[2]

이들 작업을 찬찬히 들여다보면 반복적으로 드러나 분명하게 인식되는 것이 있다. 바로 재료의 선택과 구축 과정에서 나타난 '건축의 표면' 문제다. 이들이 표면에 주목하는 이유는, 이들의 설계 방식이 프로그램 해석이나 공간을 단위로 전개하기보다는 먼저 재료를 외부 또는 내부 표면에 특수한 방식으로 사용하는 데서 출발해 보편적 건축 영역으로 확장해가기 때문이다.

표면이란 무엇인가? 건축에서 표면은 평탄한 파사드를 말하는 것이 아니며 단순히 건축 외장을 말하는 것도 아니다. 이는 스위스 건축 집단 헤르조그앤드드뫼롱 Herzog & de Meuron의 작업과 유사하다. 건물의 표피를 환경에 대응해 만든 평면적 구성 요소로 보고 의미의 근원으로 삼는다. 더 나아가 이런 건물의 통합적 표면이 건물의

내외부 환경과 관계를 맺으며 확장해간다.

건축공방의 집이자 사무실로 지어진 그들의 사옥은 큰 도로에서 떨어진 작은 골목에 있다. 이 건물의 평면은 가운데에 있는 계단실과 엘리베이터를 중심으로 대칭으로 되어 있는데 이런 질서는 상층부까지 이어진다.

외부 형태는 저층부의 기단과 상층부의 덩어리가 2층 테라스에 의해 분절된 채 들어 올려져 있다. 기단을 형성하는 저층부는 땅의 기억을 재현하듯 노출 콘크리트로 마감되어 있다. 콘크리트의 거친 인상은 좁은 수직 패턴의 거푸집으로 잔잔하고 부드러운 표면으로 전환되었다. 상층부는 2층의 일부 덩어리를 덜어내어 만든 발코니가 경계가 되어 자연스럽지만 엄격한 모듈의 아노다이징 패널로 마감되었다. 이는 자연스럽게 부식되어 진회색을 띠는데, 빛을 대부분 흡수하고 부분적으로만 반사하는 그 성질 덕분에

repeated pattern to the fore. This is the issue of the "architectural surface" highlighted through the choice of materials or the tectonic process. Their attention to the surface stems from a desire to expand our conception of architecture, by paying attention to special treatments of surface materials and not simply to develop the interpretation of the program or the spatial unit.

What is surface? The surface in architecture denotes neither a flat façade nor the simple architectural outer-skin. This is similar to the work of the Swiss architectural group Herzog & de Meuron. They see the outer skin of the building as a flat compositional element responding to the environment, taking it as the fundamental root of meaning. Moreover, the integral surface of the building progressively expands as it forms a relationship with the inner and outer environment of the building.

The headquarters built as the home and office of ArchiWorkshop is located on a small alleyway at a distance from a large boulevard. The floorplan of this building has the

stairwell and elevator symmetrically at the center, and this order continues up to the upper floors.

The external form is lifted up, with the stylobate of the lower floors and the mass of the upper floors segmented by the second floor terrace. The lower floors establishing the stylobate have been finished with exposed concrete as if to recreate the memories of the ground. The rugged impression of the concrete is converted into a serene, smooth surface through the narrow vertical patterns of the wooden molds. The upper floors are formed of natural yet strict modules of anodizing panels, bordered by the balcony carved out of the second floor. In a dark grey from natural erosion, due to its character of absorbing most light and only partially reflecting it, the joints which connects each panel are faintly perceived, establishing a non-materialized surface.

The art project <Welcome and Illusion> was based on a soon-to-be demolished house in a redevelopment district of Buk-Ahyeon-dong, Seodaemun-gu in Seoul, a dramatic

각각의 패널을 연결한 줄눈이 희미하게
인식되며 비물질화된 표면을 이룬다.

서울 서대문구 북아현동 재개발
지역의 철거될 주택을 배경으로 작업한
예술 프로젝트 ‹삶의 환영, Welcome 그리고
Illusion›은 집안을 반투명 천으로 둘러
내부화된 표면의 극적인 예를 보여준다.
오랫동안 일상을 위한 장소였으나 이제
생명을 다한 집의 벽과 개구부를 내부에서
반투명 천으로 완벽하게 감싸 마치 수의를
입힌 듯 환영幻影의 분위기를 만들어낸다.
전시가 끝난 뒤 작품에 사용한 반투명 천을
재활용해 만든 에코백을 선물로 받았는데
물질의 이동을 통해 건축공방의 집요한
의식을 엿볼 수 있었다.

‹삶의 환영, Welcome 그리고
Illusion›에는 설치 작가와 음악가와의 공동
작업을 통해 시각 효과와 음 환경을 통합한
사운드스케이프 Soundscape적 관점이 녹아
있다. 독특한 음 환경이었을 이 장소에서

바이올린과 첼로 협주가 만들어낸 1927년
파리의 격동기 음악은 어떠했을지 궁금하다.

건축공방의 작품에는 유독 임시
건축물의 일종인 ‘파빌리온’ 프로젝트가
많다. 일상의 환경에서 비일상의 시도를
하는 것으로 경계를 확장하려는 의도다.
일반적으로 파빌리온 설계는 예산이
넉넉하지 않기 때문에 주로 젊은 건축가에게
주어진다. 이는 적은 예산으로 시도하는
도전적 작업으로 흔히 기회가 되기도
하지만 무덤이 되기도 쉽다. 건축공방은
재료의 구축성과 물질성의 맥락을
변화시키는 실험적 작업들에서 줄곧 좋은
작업을 만들어냈고 그 결과물이 새로운
기회로 연결되었다. 그중 일련의 ‘글램핑
파빌리온’ 프로젝트는 스틸 프레임과
멤브레인 천으로만 제작해 로테크 low-tech 로
구현되었다. 이는 일시적 시설물이 그 한계를
극복하고, 자연을 상징하는 매개체로 높은
미학적 완성도와 함께 지속가능한 숙박

example of an internalized surface
due to the translucent fabric draped
inside the house. While the house
had served as an everyday place for
a long time, the house, now awaiting
its demise, had been completely
wrapped in the translucent fabric,
from the walls of the house to its
various exits, creating an illusory
atmosphere of being draped in
shrouds. The tenacious consciousness
of ArchiWorkshop regarding the
movement of materials was evident;
after the exhibition the eco-bag
I received was made from the same
recycled translucent fabric. Welcome
and Illusion includes a soundscape
perspective which integrates visual
effects with an auditory environment
through the collaboration of
installation artists and musicians.
It made me wonder what the height of
the turbulent era of music in 1927
Paris would have sounded like, as a
concerto of violin and cello, in this
unique auditory environment.
ArchiWorkshop's projects tend
to include many pavilions, as a
type of temporary architectural
project. These are intended as

extensions of boundaries, an attempt
at the non-everyday within everyday
environments. Normally, the design
budget for pavilions are extremely
limited, and is often delegated to
young architects. The attempt to
create with a small budget can prove
to be a challenging task and an
opportunity, but it can also become a
trap. ArchiWorkshop has consistently
produced good projects throughout
its experimental projects which
change the context of the tectonic
and qualities of the materials, and
these results have been connected
to new opportunities. Among these,
the Glamping project is a low-tech
project manufactured solely with steel
frames and membrane. It was reborn as
sustainable accommodation due to its
high aesthetic degree of completion,
by overcoming the limitations of
temporary facilities and as a medium
for symbolizing nature.
<Camptong Island & Museum> by
ArchiWorkshop near the Chunpyeong
lake of Gapyung-gun, Gyunggi-do,
originated from the characteristic
that the entire site of the lake was
made up of gneiss rock. The design

공간으로 거듭나게 했다.

경기도 가평군 청평호수 인근에 있는 ‹캠프통 아일랜드 & 미술관›은 호숫가 대지 전체가 편마암으로 되어 있다는 특징에서 비롯되었다. 설계는 대지의 속성을 고려해 두 가지 건축적 태도로 이루어졌다. 첫째는 대지의 형상과 물성에 적극적으로 반응하는 방법으로, 땅의 반향과 기억을 재현하고 대지의 속성을 변형하는 태도다. 미술관 등의 시설은 기하학적 단순성을 갖지만 땅의 형상을 은유적으로 재현한다. 둘째는 마치 나무처럼 최소한의 면적으로 대지에 구축되어 있어 땅을 존중하고 프레임과 중성적인 감각의 멤브레인 천으로 만들어졌다. 두 재료와 구축법의 대비는 자연에 대한 존중과 건축의 존재감을 명료하게 드러내고 사용자가 특별한 경험을 할 수 있게 한다.

"좋은 건축은 심미성과 기능성의 조화에서 비롯된다. 미학과 기능의 균형을 잡는 일이다. 이는 건축 분야의 협업에서 밸런스를 만드는 것이기도 하다."[3]

비물질성은 하드웨어로서 건물의 가치를 다양한 문화 현상을 담는 그릇의 차원으로 격상하는 데 기여한다. 일상의 공간은 이미 경험한 세계이므로, 직관적 판단만으로도 건축물로 구성했을 때의 물질의 속성과 비물질적 특성에 대한 높은 수준의 사전 예측과 미학적 판단이 가능하다. 그러나 비일상성의 세계는 기능 또는 미학적으로 경험해보지 못한 영역이므로 직관만으로는 불충분하다. 새로운 실험을 건축 안으로 끌어들이기 위해서는 관념적 태도가 요구된다고 하겠다.

considered the qualities of the site, and took two architectural approaches. The first was a method of actively responding to the form and materiality of the site, in an attempt to recreate the reverberations and memories of the land to mold the characters of the site. While facilities like the museum possess a geometrical simplicity, they metaphorically recreate the form of the land. Secondly, it is built on the ground with a minimum area like a tree. It is made of membrane with respect to the ground and a frame and neutral sense. The contrast between the two materials and construction methods reveals respect for nature and the presence of architecture and allows the user to have a unique experience.

## FROM INSTINCT TO CONCEPT

"Good architecture originates from the harmony of aesthetic form and functionality. This involves forming an equilibrium between aesthetics and function. This also creates a balance in collaborations taking place within the field of architecture."[3]

Immateriality contributes to elevating the the value of a building as the hardware for the dimension of a vessel, which embraces a variety of cultural phenomena. Everyday space is the already experienced world, whereby intuitive judgment can permit a high standard of anticipation and aesthetic judgment about its material qualities and immaterial character when a building has been composed. However, the extra-ordinary world is a domain unexperienced through neither function nor aesthetics, and hence cannot be met with intuition. We could say that a conceptual approach is required to introduce new experiments into the architecture.

To Edward Husserl, the German philosopher, the everyday was an intuitive world of experience, contrasting with the scientific and conceptual structures. While the everyday can be intuitively articulated and analyzed, in contrast, the extra-ordinary is featured as

독일 철학자 에드문트 후설Edmund Husserl에게 일상은 무엇보다도 과학적이고 관념적인 구성체와 대비되는 직관적 경험의 세계다. 일상성은 구체적으로 직관할 수 있고 파악할 수 있지만, 이에 반해 비일상성은 관념적 구성체를 나타낸다. 따라서 일상성은 특별한 전문적 작업 과정을 개입하지 않고도 자유롭게 처리할 수 있다. 하지만 비일상성은 인위적 규칙에 따르며 전문적 지식 체계를 요구한다. 일상은 무문제성無問題性 신뢰성의 기초를 이루며, 비일상성은 불확정적이고 끊임없이 검토된다.

건축공방이 바라보는 일상성은 르페브르의 인식처럼 안정을 추구하는 곳이라기보다는 변화를 전제로 하는 장이다. 변화의 추구는 비일상적이다. 이는 관념적 구성체로 나타내기 위한 전문적이면서도 새로운 전략이 필요하다. 건축공방의 향후 과제는, 재료와 표면으로부터 전개되어 온 관심이 물리적으로 구체화될 때 비일상성을

다루는 관념의 영역과도 조우하는 건축 방법론을 찾는 것 아닐까.

건축공방의 박수정, 심희준은 아틀리에 건축공방의 공방을 '攻防' 혹은 '工房'이라고 해석한다. 두 건축가는 매번 논쟁에 가까운 토론을 거쳐 규모가 큰 건축 작업이든 규모가 작은 소품 디자인든 상관없이 프로젝트마다 하나씩 고유 번호를 부여하며 완성도 높은 결과물로 끌어내고자 한다.

건축가마다 독립된 사유를 한다는 점을 고려하면 건축이 협력이 아닌 공동의 산물이기 어렵다는 견해도 있다. 파트너십을 통한 작업일 때도 주 역할과 부 역할이 나누어져야 한다고도 한다. 하지만 두 건축가는 자신들의 작업에 관해 설명할 때 역할을 구분하지 않고 항상 나누어 설명한다. 필자는 건축이 한 건축가의 고유한 작업일 수밖에 없다는 작가주의적 견해보다는 고유함이든 혁신성이든 공동 감각을

conceptual structures. The everyday can be freely dealt with without the intervention of a particular expert work process. However, the extra-ordinary must follow artificial rules, and requires expert knowledge systems. The everyday is based on a trust of unquestioning stability (無問題性), while the extra-ordinary is uncertain and continuously under review. The everyday as seen by ArchiWorkshop is a place based on the presumption of change, rather than a place to pursue stability as Lefebvre perceived. This requires strategies that are simultaneously expert and new to illustrate this as a conceptual structure. Perhaps the future task of ArchiWorkshop will be to seek out possibilities for an architecture, materially articulated through the development of the materials and the surface, which meet the conceptual realm where one can deal with the extra-ordinary nature of architecture.

Sujeong Park and Heejun Sim of ArchiWorkshop interpret the Korean term "workshop" (*gong-bang*) in two senses: as to construct a defense/

fortress (攻防) or as an artisanal workshop (工房). With each project, the two architects attribute a system of numbers to each project, whether a large-scale architectural project, or a small-scale product design, administering a process of debate and discussion each time to produce a high-level of completion.

Considering that each architect possesses their own capacity for thought, perhaps architecture as a common product is unfeasible, unless it is through collaboration. The architects say that they split the roles in terms of main and sub-responsibility, even for projects worked in partnership. However, the two architects have divided rather than distinguished their roles when describing their work. As the critic, I highly value objects composed through a common decision making process based on a sense of the unique, or the innovative or a common sensibility, rather than the Authorship viewpoint that architecture cannot help but be the unique work of a single architect. It seems that the two architects of ArchiWorkshop

바탕으로 한 공통의 의사 결정 과정에 따라
만들어지는 사물의 가치에 거는 기대가
더 크다. 건축에서 중요한 것이 고유성과
함께 보편성을 획득하는 과정이라고 할
때, 건축공방의 두 건축가는 좋은 선택지를
갖고 있는 셈이다. 이미 성숙한 작업을
하고 있는 두 건축가가 '2019 젊은건축가상'
수상을 계기로 사유의 지평이 넓어지길,
사회에서 새로움을 시도하고 의미 있는
작업을 지속하는 건축가 그룹으로 성장하길
기대한다.

have a good deck of cards to pick
from, if one were to consider the
most critical part of architecture
as the process of acquiring a sense
of uniqueness. I hope that the two
architects, who already produce a
mature level of work, gain momentum
from being awarded the 2019 Korean
Young Architect Award, growing as
an architect group that widens its
horizons of thought, experiments anew
in society, and continues to work on
meaningful projects.

에필로그　　　　　　　2019 젊은건축가상
　　　　　　　　　　심사 총평
　　　　　　　　　　: 김헌

Epilogue　　　　　　　The 2019 Korean
　　　　　　　　　　Young Architect
　　　　　　　　　　Award - Evaluation
　　　　　　　　　　Review
　　　　　　　　　　: Hun Kim

김헌
2019 젊은건축가상 심사위원,
스튜디오 어싸일럼 대표

사실 연령의 차이만 다소 있을 뿐 현재
같은 시공간에서 함께 뛰고 있는 후배
건축인들의 작업을 심사하는 일이 내겐 늘
어색할 수밖에 없다. 우리 현실, 우리 건축의
특성상 무슨 일에서든 그 내면에 어떤
장애와 수준의 한계가 도사리고 있었는지
짐작하고도 남는 마당에 짐짓 모른 척 경직된
잣대부터 들이밀기가 영 내키지 않아서다.
그런 탓인지 다른 사람은 모르나 내 경우
'젊은건축가상'과 같은 이벤트는 딱히
비평이나 평가의 행위라기보다 작품들에
대한 흥미, 호기심, 감상에 젖게 하는 일에
더 가깝다. 그 끝에 남다른 성원이나 격려의
표를 다소 더 얻은 팀이 수면에 떠 오르는,
하나의 인기투표 같은 절차라 할까.

한편으로 어쩐 일인지 '2019
젊은건축가상'은 과거 어느 해보다 확연히
많은 마흔세 개 팀이란 지원자 수를
기록했다. 관계자들이 추측하기를 올해
새로이 심사를 맡은 인물들의 면면이 그

Hun Kim
Jury of 2019 Korean Young Architect Award,
Principal of Studio asylum

I can't help but feel a slightly
apprehensive when judging the work of
up and coming architects. In truth,
as architects living and breathing
in the same space and time, the
only thing that sets us apart is
our age. I feel reluctant to impose
some sort of rigid framework with
which to judge their work, when we
are all well aware of the inherent
difficulties and existing limitations
enforced by the given reality and
contemporary conditions of Korean
architecture. Perhaps this is why,
for me at least, events like the "Young
Architect Award" is less an act of
critiquing or evaluating the work of
others, and instead are significant
as moments in which to engage others
in one's impressions, interests,
and fascination with certain
architectural projects. Perhaps it is
best described as a sort of popular
vote, where the teams who have
elicited more votes of support or
encouragement surface above the rest.

In one sense, the 2019 Korean
Young Architect Award caught us by
surprise, as it had a record number
of applications, with forty-three

숫자에 나름 크게 영향을 미친 것은 아닌가 한다지만 그 역시 과연 어떤 연관성이 있는 것인지는 모르겠다. 응모 팀이 급증했다는 건 그만큼 전례 없이 방대한 검토 자료가 우리 몫으로 던져졌다는 것을 뜻하기도 한다. 게다가 응모한 각 팀의 작업 역시 그 성과의 밀도라든가 특징적인 면에서 고루 우수하고 일견 큰 차이가 보이지 않아 2차 공개 심사를 위한 여덟 개 팀을 가리는 일조차 나에게는 난항이었다. 자연히 공개 심사 후 최종 세 개 팀을 수상자로 선정하는 단계에서 오랜 시간의 논의와 숙고가 불가피했다.

매해 그랬을 터이지만 이런 경우 역시 관건이 되는 것은 심사위원들 각자의 현역다운 강한 개성과 인식의 차이다. 초반 얼마 동안은 그런 점들을 서로 의식하면서 과연 의견의 수렴이란 것이 이루어지기는 할까 하는 관망이 없지 않았다. 물론 시간이 흐르면서 아슬아슬하지만, 개개의 톱니들이 맞물려가던 순간들을 차츰 맞이하기는

했지만 말이다. 참고로 내가 애초에 공들여 표를 던졌던 팀들은 심사가 거듭되면서 마치 손가락 사이로 모래가 빠져나가듯 남김없이 모두 힘을 잃고 쓸려나갔다. 그런데 따지고 보니 올해 수상자인 세 팀 중 두 팀 역시 과거에 한 번 이상 응모했던 이력을 갖고 있지 않은가. 그들도 그때는 선택받지 못했던 팀들의 어두운 심정이었으리라.

최종 수상자들 중 아이디알의 이승환과 전보림의 작업들은 공공 건축 분야가 본질적으로 지닐 수밖에 없는 수많은 제약과 한계에 맞서 분투한 상흔傷痕들을 점점이 새기고 있다. 관제 건축 본연의 인색한 어젠다agenda, 그 표면에 꼼꼼히 영특한 틈새들을 벌려 무미건조함이나 진부함을 극적으로 돌파하는 다채로운 시나리오가 그 배경에 깔려 있다.

또 다른 팀, 건축공방의 박수정과 심희준 두 사람은 올해 여타 지원자보다도 현저하게 확장된 직능의 스펙트럼을

submissions. While the organizers have hazarded the opinion that this year's assortment of characters on the review panel may be the main instigator of such enthusiasm, no substantial claim exists to support this idea. The acute upturn in candidates also meant that an unprecedented quantity of material was given to us for review. Moreover, the uniformly exceptional quality of the work and character of the projects, and the density of each team's accomplishments, made it difficult to select just eight teams to make it to the second round of open adjudication. Naturally, this round was inevitably followed by a long process of discussion and consideration, as it was necessary to nominate three teams as the final winners.

While this must be a recurring issue every year, this year it was the strong characters and different views of the judges that was key to identifying the winners. The early stages of the review involved a heightened sense of awareness about each other, accompanied by a wait and see attitude as to whether a true

consensus could really be reached. Of course, despite some touch-and-go moments, with the passing of time, moments started to emerge where individual cogs started to click into motion. I would also like to mention that the teams which I had judiciously selected at the onset eventually all slipped away like loose sand between my fingers. Then again, at least two teams out of the three final winners have a track record of having already applied as candidates to this award more than once. These teams would have also experienced the shadows currently cast over the minds of the unselected.

Among the final winners, the work of Seunghwan Lee and Borim Jun of IDR are engraved with the scars of those who have fought against the countless conditions and limitations inherent in the field of public architecture. Their work is based on lively scenarios which dramatically break through the mundanity and insipidity which is too often a characteristic of stingy government-commissioned architecture, by shrewdly discovering and drilling down on the

드러냈다. 거시적 규모의 도시 재생을 다룬
구상에서부터 랜드 아트, 설치 미술, 환경
조형 작업, 게다가 모바일 구조물, 가구 등
산업 디자인의 영역마저 파고들며 이 넓고
다양한 분야를 거침없이 오가는 모습은 거의
독보적이다. 물론 첨예한 현실 속에서 법적
기술적 매듭과 난제를 효율적으로 풀어내고
있는 보편적 일상의 건축 행위는 차치하고도
말이다.

끝으로 푸하하하프렌즈의 작업에서는
우리 시대 우리 건축 수행 방식의 한 가지
고유한 유형을 건져낼 수 있었다. 또 그것이
앞으로도 한동안은 쉽게 나타날 것 같지는
않을 내용과 특성이란 면에서 다수의 표를
얻게 한 것 같다. 그들의 건축에서는 이른바
엘리트 의식이란 것을 먼저 배제하고 애초에
어떤 차용이나 복제의 뜻이 보이지 않으면서
단지 장소나 주어진 여건에 근거한 해법을
펼치고 있다는 흔적이 역력했다는 후기다.

당연한 일이지만 비록 올해에

한정한다 해도 이 상의 취지와 관련해 심사
위원들끼리 서로의 공감을 확인하는 시간을
사전에 가진 바 있다. 즉 '젊은건축가상'이란
단순히 누군가의 성과물을 평가하고
치하한다는 좁은 시야의 의미에 머물러서는
안 되며 2019년을 사는 모든 건축인에게
어떤 메시지를 발하는 하나의 반영체로
기록되어야 한다는 점이다. 험준한 시간과
공간 속에서 함께하며 이른바 분투라 불리는
작가적 일상을 사는 사람들에게 상정하는
메시지일 터이다.

응모 작업을 살피는 과정에서 그간
우리가 유독 높이 산 점들, 결과물을 보며
아낌없는 독려의 말을 건네고 싶은 면면들,
또 과연 우리가 생각하는 건축의 진정한
참신함이란 어떤 것인지 같은 그런 맥락에서
올해의 수상자들, 또 더욱 앞서 곡절 끝에
이번 공개 심사에 최종적으로 서게 된 여덟
개 팀, 모두를 폭넓게 관통하고 있는 특질 몇
가지가 자연스레 드러나고 있었다.

274

smallest crevices of possibility.
Another team, Sujeong Park and
Heejun Sim of ArchiWorkshop explicitly
features a more extensive spectrum
of work than any other candidate.
The pair were singular in their
unrelenting ability to shift from
genre to genre, from their conception
of macro-scale urban regeneration
to land art, installation art,
environmental sculpture, and even the
field of industrial design such as
mobile structures and furniture. This
is, of course, through the execution
of a ubiquitous everyday architecture
which effectively resolves legal and
technical complications and challenges
amidst a conflicting reality.
Finally, the work of FHHH
friends allows us to extricate
a single unique typology in how
architecture is practiced in this
generation. It seems that the multiple
votes were in favor of its content
and characteristics, which we agreed
may be difficult to glimpse again in
the near future. Their architecture
first eliminates any trace of elitist
architecture, and designs solutions by
drawing solely from a locale or from

given conditions, without resorting
to methods of irresponsible or
thoughtless adoption or replication.
As is the most logical thing
to do, the adjudicators got together
before the competition to affirm
a common understanding regarding
the ultimate goal of this award. We
agreed that the Young Architect Award
must not settle upon a narrow field
of vision which simply evaluates and
commends  accomplishments, but that
it should go further to record as a
single reflective body a message to
transmit to all architects living in
2019. This would be a message for
those who live everyday as an artist,
fighting the so-called virtuous
fight, by co-existing in this
perilous present time and space.
When reviewing the submissions,
and considering the values that we
felt strongly about, the pieces
we wished to encourage, and more
abstractly, representative of
authentic and creative architecture,
we found that the candidate portfolios
revealed several widely permeating
characteristics not only common to
the work of this year's winners, but

우선 이 첨예한 자본 중심의 시대에, 어느덧 가격price이 곧 가치value라고 들뜨듯 열창하고 있는 현실에서 이들의 포트폴리오를 채우고 있는 것들에는 초미에 번들거리는 기름기부터 걷어낸 형상과 물성의 렌더링이 지배적이었다. 어느 챕터를 보아도 한결같이 간소하나 영리한 언어들이 열거되어 있는 다큐멘터리이기도 했다. 게다가 건축이 개입intervention하면서 품게 된 지역이나 사회, 사용자 그룹에 대한 깊이 있는 애정 내지는 배려 의식이 일관성을 잃지 않고 있었다. 발상의 서두는 지극히 사적인 이론이나 개념이었을지라도 이것이 어떻게든 온기를 지닌 채 타자와 치열한 관계 맺기에 이른다. 나아가 자칫 치기 어리기 쉬운 특유의 작가 의식에 대해, 때로는 자기 복제에 대해 스스로에게 가하는 엄한 경계와 단속 또한 공히 감지되는 것 같아 반가웠다. 따라서 『젊은 건축가: 질색, 불만 그리고 일상』의 각 서두에 새겨진 각 수상자의 이름과 영예

뒤에 훨씬 더 휘황찬란한 메시지 보드가 점멸하고 있음을 살펴보기 바란다. 애초부터 그렇듯 '젊은건축가상'의 상賞은 사실 상像이었음을 밝힌다는 것을 말이다.

also the eight teams presenting at this year's open round adjudication.

First, in this acutely capitalist generation, and amidst a reality in which somehow price is fervently equated to value, the portfolios of these architects were first and foremost filled with renderings that featured forms and materiality which, in the main, avoided extravagant excess. Each chapter seemed to be composed like a documentary, connected by simple but insightful words. Moreover, they consistently pursued a deep sense of affection or consideration towards local communities, society or user groups encountered through their architectural intervention. While the inception of the idea may originate from personally meaningful theories or concepts, these connections with others are formed with genuine warmth. Furthermore, it was heartening to see a strict sense of self-awareness and self-regulation regarding that particular self-awareness as an artist, which can all too often be mischaracterized as a trait of youth, or at times self-replication. Hence, as you read 『Young Architect: Loathing, Dissatisfaction and the Everyday』, our hope is that you view each of the winners named at the beginning of each of the chapters in the context of this energized and momentous conversation. With this, we hope to declare, as it always has been, that the Young Architect award is truly also the image of Young Architecture.

푸하하프렌즈 크리틱 참고문헌

1   윌리엄 모리스, 『지상낙원』,
    「두 번 다시 웃지 않는 사나이」, 1870.

2   가스통 바슐라르, 정영란 옮김,
    『공기와 꿈』, 이학사, 2000.

3   존 러스킨, 현미정 옮김, 『건축의 일곱
    등불』, 마로니에북스, 2012.

4   히토 슈타이얼, 안규철 옮김,
    『진실의 색』, 워크룸 프레스, 2019.

5   장 뤽 고다르 감독, ‹필름 소셜리즘›,
    2010.

6   로완 무어, 이재영 옮김, 『우리가 집을
    짓는 10가지 이유』, 도서출판 계단,
    2014.

7   로완 무어, 이재영 옮김, 『우리가 집을
    짓는 10가지 이유』, 도서출판 계단,
    2014.

8   크리스토퍼 알렉산더, 한진영 옮김,
    『영원의 건축』, 안그라픽스, 2013.

9   로완 무어, 이재영 옮김, 『우리가
    집을 짓는 10가지 이유』, 도서출판
    계단, 2014.

FHHH friends'Critique Reference

1   William Morris, 『The Earthly
    Paradise』, 「The Man Who Never
    Laughed Again」, 1870.

2   Gaston Bachelard, translated
    by Youngran Jeong, 『Air and
    dreams』, Ehaksa, 2000.

3   John Ruskin, translated by
    Mijeong Hyun, 『The Seven Lamps
    of Architecture』, Maroniebooks,
    2012.

4   Translator's note: Pronounced
    Poohahaha Friends in Korean

5   Hito Steyerl, translated by
    Kyuchul Ahn, 『Die Farbe der
    Wahrheit』, Workroom Press, 2019.

6   Translator's note: To one's
    literary debut officially
    recognized

7   Director Jean-Luc Godard, <Film
    Socialisme>, 2010.

8   Rowan Moore, translated by
    Jaeyoung Lee, 『Why We Build』,
    Publisher Gyedan, 2014.

9   Rowan Moore, translated by
    Jaeyoung Lee, 『Why We Build』,
    Publisher Gyedan, 2014.

10  Christopher Alexander,
    translated by Jinyoung Han,
    『The Timeless Way of Building』,
    agbook, 2013.

11  Rowan Moore, translated by
    Jaeyoung Lee, 『Why We Build』,
    Publisher Gyedan, 2014.

아이디알 크리틱 참고문헌

1       IDR Architects 블로그, 2018년 6월
        19일 게재, https://blog.naver.com/
        idrarchi/221302333584.

2       같은 블로그, 2018년 10월
        20일 게재, https://blog.naver.com/
        idrarchi/221381196038.

3       같은 블로그, 2019년 6월
        12일 게재, https://blog.naver.com/
        idrarchi/221560346441.

4       같은 블로그, 2018년 10월
        20일 게재, https://blog.naver.com/
        idrarchi/221381196038.

IDR's Critique Reference

1       IDR Architects blog,
        published the 19th of June,
        2018, https://blog.naver.com/
        idrarchi/221302333584.

2       Same blog, published
        the 20th of October,
        2018, https://blog.naver.com/
        idrarchi/221381196038.

3       Same blog, published the 12th of
        June, 2019, https://blog.naver.
        com/idrarchi/221560346441.

4       Translator's note:
        To subject oneself to hardship
        of all kinds in order to
        strengthen one's resolution to
        wipe out a national humiliation
        or to attain an ambition.

5       Translator's note:
        To return in a swirl of dust,
        to make a stage comeback, to
        rebound.

6       Same blog, published
        the 20th of October,
        2018, https://blog.naver.com/
        idrarchi/221381196038.

건축공방 크리틱 참고문헌

1    «건축문화» 「작가특집-건축공방」,
     건축공방 글, 2019년 2월호, 84쪽


2    같은 책, 85쪽


3    같은 책, 84쪽

ArchiWorkshop's Critique
Reference

1    «Architecture and Culture»,
     「Architect Series —
     ArchiWorkshop」, written by
     ArchiWorkshop, February 2019
     issue, p.84

2    same book, p.85

3    same book, p.84